# Fuzzy Classifiers
# and their Relation to Cluster Analysis and Neural Networks

Birka von Schmidt

Von der Carl-Friedrich-Gau-Fakultät für Mathematik und Informatik
der Technischen Universität Braunschweig genehmigte Dissertation zur
Erlangung des Grades einer Doktorin der Naturwissenschaften (Dr. rer. nat.)

Bibliografische Information der Deutschen Nationalbibliothek

Die Deutsche Nationalbibliothek verzeichnet diese Publikation in der
Deutschen Nationalbibliografie; detaillierte bibliografische Daten sind
im Internet über http://dnb.d-nb.de abrufbar.

ISBN 978-3-8325-1463-1

Logos Verlag Berlin
Comeniushof, Gubener Str. 47,
10243 Berlin
Tel.: +49 030 42 85 10 90
Fax: +49 030 42 85 10 92
INTERNET: http://www.logos-verlag.de

# Acknowledgments

Many people have contributed to this book in one or the other way.

First of all I would like to thank Prof. Dr. Frank Klawonn and Dr. Johannes Reichmuth. When the idea came up to use Neural Networks for the optimization of flight arrival times, it was only a vague idea. I would like to acknowledge Prof. Dr. Klawonn for his support in turning this idea into a theory, for the intense discussions and his valuable input on content and presentation format of the results. Dr. Reichmuth earns my gratitude for his professional and personal support during the phases of theoretical work, implementation and application of the results on optimization problems for flight arrivals.

I would also like to thank the other member of the Department of Flight guidance systems at the German Aerospace Center for their support. Without the German Aerospace Center as such, this thesis would not have been possible.

I owe special thanks to Dr. Annette Temme for the discussions about fuzzy clustering methodology, as well as for her motivating me. Frank Rehm helped me by providing input on weather data and fuzzy clustering of these data. The neural network simulations of this book were carried out by software from Dr. Christian Borgelt. I would also like to thank Prof. Dr. Rudolf Kruse for his input in the course of the Working group for Softcomputing and for taking the effort as a reviewer for my thesis.

Last but not least I would like to thank Prof. Dr. Hans Opolka for setting the foundation for my scientific work, and my family and friends for their respect, support and encouragement during the years I spent with this thesis.

4

# Contents

5

# Chapter 1

# Introduction

As classical computing methods use mathematical methods founded upon the distinction between 0 and 1, between *true* and *false*, between the two ends of a scale, their applications fields are restricted. They can rarely deal with incomplete or imprecise data.

Since fuzzy sets where introduced in 1965 [107], new methods have found their way into computational methodologies and data analysis [8]. Fuzzy logic [22, 63] and fuzzy rules tried to model a way of calculating with incomplete or imprecise data and thus offered new possibilities [20]. Other techniques as evolutionary or neural computation are approaches based on learning rules from existing data.

The field of soft computing includes different methodologies, as e.g. rule based fuzzy classification systems or fuzzy clustering as well as neural networks. These three are the methods we will examine in detail, although there exist others.

Focusing on fuzzy systems of any kind and analyzing them leads us to a deeper understanding of internal procedures. This knowledge enables us to visualize the processes transpiring when the systems assigns an output to the data. And the latter opens the view on possibilities to improve these systems, either in their performance or in their interpretability.

From a theoretical point of view fuzzy controllers, that are more common than fuzzy classifiers, are a method to describe a real function $\mathbb{R}^m \to \mathbb{R}$ (or, in the case of multi-input, multi-output systems, $\mathbb{R}^m \to \mathbb{R}^k$) assigning a real (control) value to a given tuple, point, or vector of measured input values. There is a variety of different models of fuzzy controllers like the Mamdani-type controller [65] that uses fuzzy sets in the consequent part of the rules or the Takagi-Sugeno model [101] that allows a (linear) function of the inputs in the consequent part of the rule. For an overview see for instance [54].

In almost all fuzzy control systems, the final crisp output is computed incorporating the outputs of all rules whose premises are satisfied to a degree greater than zero, i.e. whose firing degree is greater than zero. There are many different ways to aggregate the outputs of the single rules and – in the case of a Mamdani controller – to defuzzify the resulting fuzzy set. Nevertheless, the underlying principle is always that the output is some kind of weighted mean of the outputs of the firing rules.

Fuzzy controllers are well examined as function approximators. Piecewise monotone functions of one variable can be exactly reproduced by a fuzzy controller [5, 62] and for the multi-dimensional case fuzzy controllers are known to be universal approximators for continuous functions [15, 50, 105]. However, these positive general results do not apply when the number of rules is restricted. In this case, the set of functions that can be represented by a fuzzy controller is nowhere dense [74].

These results do not apply to fuzzy classification systems. The situation is different, since we have to deal with a function $\mathbb{R}^m \to \mathcal{C}$ where $\mathcal{C}$ is a finite set of discrete classes. Fuzzy classification systems differ from fuzzy controllers in the form of their outputs. For classification problems a decision between a finite number of discrete classes has to be made, whereas in fuzzy control the output domain is usually continuous, i.e. an interval of real values. We do not assume any kind of structure on $\mathcal{C}$. This means that interpolation between classes does not make any sense.

The classes could for instance be different runways to which the arriving aircrafts have to be assigned by the air traffic control, or the gates where passengers have to change flights.

Fuzzy classification systems of this type have been successfully applied to a number of problems (see for instance [31, 38, 47, 68, 75, 106]), but a systematic experimental or theoretical analysis of these systems was initiated just recently.

This work will concentrate on classifiers, setting off by analyzing the Max-Min-Fuzzy classifier in chapter 3 after a short introduction to fuzzy classifiers in chapter 2. Using the maximum and the minimum to calculate the classification provides us with a well known and commonly used rule based fuzzy classification system. Examining fuzzy max-min classifiers in chapter 3 opens us insight into the system and reveals certain restrictions. It becomes obvious that a max-min fuzzy classification system is not able to describe an arbitrary classification, because it decides locally on the basis of two attributes.

Graphically spoken, this means that the separation described by the classifier is drawn by separating hyperplanes in the data space that are except for two dimensions parallel to the coordinates. Only two attributes influence the position of the plane.

As this strong restriction handicaps the construction of appropriate classifiers, we search for a classifier with a wider range of application for more complex classification problems. When using the Łukasiewicz-t-norm and -conorm instead of minimum and maximum in chapter 4, we obtain a solution.

We analyze the Łukasiewicz-t-norm and -conorm and visualize the classifications described by such a system. These norms provide us with the means to define a classification system that is competent to solve arbitrary linearly separable problems.

The main principle of the construction is that the space is divided into smaller cuboids that include only hyperplanes of a geometry as simple as possible. Setting up rules for each of these cuboids, we can compound the resulting rules to a classification system covering the whole data space. We will give detailed constructions in chapter 5.

In the beginning we mentioned different methods of soft computing that can be used for similar problems. Each of these methods has it's advantages and disadvantages. Thus a combination of different systems would reveal new facilities and diminish the disadvantages of a single system. Later we will suggest different approaches for such combined systems, where the basic technique is nearly always the same. It relys on the visualization of the classification via separating hyperplanes in the data space.

Fuzzy clustering which we will briefly review in chapter 6 is known as a method for combining data with similar attributes into groups. Although fuzzy clustering usually is applied in the case of unsupervised classification with the data from the training set not being labeled, there are also methods to use fuzzy clustering for supervised classification [104].

The most common fuzzy clustering method is the fuzzy c-means clustering. Each cluster of similar data is represented by a prototype and the data vectors are assigned to the cluster represented by the prototype that is closest to a data vector. This leads us again to the visualization via hyperplanes. When we draw a plane in the middle between two prototypes, then the data is assigned to the prototype that is on the same side of the separating hyperplane.

Since a classifier derived from fuzzy clustering and a rule based fuzzy classifier both use multi-dimensional membership functions, the first is often transformed into the latter, mainly because a rule-based fuzzy classifier is easily interpretable. A common means for this transformation is projection, although this often means a loss of information, while a decision algorithm based on hyperplanes drawn through the data space can reproduce the information of the clusters exactly.

Therefore in chapter 7, we visualize the fuzzy clusters via separating hyperplanes. These hyperplanes can be the starting point for either the construction of a rule based fuzzy classification system or the multilayer perceptron, both performing the classification done by fuzzy clustering.

As well as we are able to derive rules from fuzzy clustering results, other research was done on deriving fuzzy rules from decision trees. [81, 33, 86, 39] The multilayer perceptron (MLP) is an artifical neural network. Its principles were motivated by mechanisms in neurons of the human brain, enabling the MLP to learn and to improve it's performance. It has a good performance and learns quickly and can therefore be used for a wide range of problems. But it is a 'Black Box', because it is difficult and is often impossible to interpret what happened inside.

Other models of artificial neural networks as e.g. the radial basis function networks or Kohonen's networks have been successfully interpreted, but the MLP still waits for useful solutions.[108, 28]

In the classical definition of the MLP the activations of the units and therefore the output spaces are continuous. The sigmoidal activation function enables the MLP to learn by using the backpropagation algorithm.

As we use the MLP for classification there is no reason for interpolation between the classes [88]. We aim at distinct outputs. We can achieve e.g. this by choosing the activation function in an appropriate way, i.e. as steep as possible and restricting the output to 0 and 1. Nevertheless, the activation function has to be less steep in the beginning of the learning process to secure learning progress.

The MLP has the characteristic that its calculations can be considered as defining hyperplanes in the data space, which enables us to draw parallels between the fuzzy classification system and the MLP.

After a short introduction to MLPs in chapter 8, the analysis of these hyperplanes in chapter 9 leads to a method to transform a fuzzy classification system into a MLP. The resulting MLP describes nearly the same classification. The results are slightly different due to the different calculation methods the two system use.

Other approaches to construct neural networks from cluster data can be found in the literature, see e.g. [16].

The transformation of one system into the other gives us some advantages, that become obvious when we compare the differences between the two: The fuzzy classification system is interpretable, while the multilayer perceptron can improve it's performance by learning.

Imagine a number of data sets, e.g. measured data over a period of time. There are relatively small changes from a precedent data set to the following, but the data sets themselves are huge and difficult to handle. When we can set up a fuzzy classification system for the first of these sets, we can construct an MLP that represents nearly the same information as the fuzzy classification system.

For the following periods we do not have to repeat the whole process, but we can rely on the MLP for the first year and let it adopt to the new data set by performing the learning process.

Such am MLP can be used e.g. as a mean for prediction. A challenge for data analysis is the field of air traffic with it's problem of delays in arrival times of aircrafts. There are still many factors to be examined that could influence the delay times. Available runway capacity, traffic and organizational factors are just a few to mention. Any improvement of knowledge about the influence of these factors helps the air transport stakeholders to plan and manage the air traffic in a more efficient way. In chapter 10 we examine weather data.

There has been a lot of research in combining different methods into one system, e.g. using fuzzy techniques in neural networks [75, 49, 48, 70], some even concentrate on classification problems [82, 67, 69, 29, 51]. There has been work on combining different classifiers [61, 59, 60, 56] but these are different approaches towards the combination of their advantages than the one we follow. Other approaches for deriving fuzzy rules from neural networks can be found in [14, 1, 23, 103, 71, 72].

We examine the three systems, rule-based fuzzy classification systems, fuzzy clustering and multilayer perceptron, separately and uses their common geometry to establish methods to transform one into the other while preserving the geometric characteristics.

Additional ways of visualizing fuzzy data and neural networks can be found in [9, 30, 45].

# Chapter 2

# Fuzzy Classifiers

Fuzzy classification systems model a function $f : \mathbb{R}^m \to \mathcal{C}$ where $\mathcal{C}$ is a finite set of classes.

Nürnberger et al. [79] investigated the class boundaries of two- and three-dimensional data that can be generated by fuzzy classification systems using different t-norms. A short overview will be given in section 2.2. Cordón et al. [17] analyzed fuzzy classification systems on an experimental basis that do not rely on a classification based on the rule that best fits the input. L. Kuncheva provides a very thorough and detailed analysis of fuzzy classifiers in [57].

In this chapter we will introduce fuzzy classification systems and their characteristics (for other classification methods see e.g. [40, 12]).

## 2.1 Description of Fuzzy Classifiers

In this section we will present the formalized definitions of fuzzy classification systems. We start with the description of fuzzy sets and continue until we have defined a fuzzy classification set.

### 2.1.1 Fuzzy Sets

**Definition 1**
*Let $X \neq 0$ a set. Then a fuzzy set on $X$ is a function*

$$\mu : X \to [0; 1].$$

*The set of all fuzzy sets is denoted $F(X) := \{\mu \mid \mu : X \to [0; 1]\}$.*

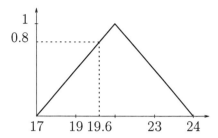

Figure 2.1: The fuzzy set for 'approximately 21°C'.

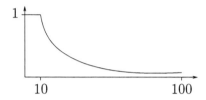

Figure 2.2: The fuzzy set for 'young'.

**Example 1**

*For the climatisation of a room the temperature has to be controlled. The fuzzy set in figure 2.1 describes the linguistic value 'approximately 21 degrees Celsius.*

**Example 2**

*A number of people aged between 10 and 100 are asked whether they feel young. We define A to be the set of people that are interviewed and B :=* $\{1, \ldots, 100\}$. *The function*

$$y : B \to \{1, \ldots, |A|\}$$

*assigns to each $b \in B$ the number of people that have this age and feel young, while $n(b)$ is the number of people that are aged b. Then the function*

$$\mu : B \to [0, 1] \text{ with } \mu(b) = \begin{cases} 1 & \text{if } b > 10 \\ \frac{y(b)}{n(b)} & \text{if } y \geq 10 \end{cases}$$

*is the fuzzy set that describes the age of feeling young. As this example does not deal with a vague object, but a with the vagueness of description, it has to be used a possibilistic interpretation on this problem.*

It is important always to keep balance between the interpretability, that we achieve by transforming linguistic terms into fuzzy sets and the accuracy.

Assume that we have a fuzzy set for describing whether a person is *young* and one for describing he is *tall*. When we want to determine whether a person is *young and tall*, we need an conjunction of the two fuzzy sets. In the case of fuzzy sets the conjunction is determined by the *t-conorm*.

**Definition 2**
*A function* $\top : [0, 1] \times [0, 1] \rightarrow [0, 1]$ *is called a t-norm iff it fulfills*

1. *For all* $a \in [0, 1]$ *the equation* $\top(a, 1) = a$ *holds. (1 is the unitary element.)*

2. *For all* $a, b, c \in [0, 1]$ *the equation* $a \leq b \Rightarrow \top(a, c) \leq \top(b, c)$ *is fulfilled. ($\top$ is monotonously not decreasing.)*

3. *For all* $a, b \in [0, 1]$ *the equation* $\top(a, b) = \top(b, a)$ *holds. (Commutativity)*

4. *For all* $a, b, c \in [0, 1]$ *the equation* $\top(a, \top(b, c)) = \top(\top(a, b), c)$ *is fulfilled. (Associativity)*

Finally we need a function to define the disjunction. The disjunction is determined by the *t-conorm* as defined in the following definition.

**Definition 3**
*A function* $\perp : [0, 1] \times [0, 1] \rightarrow [0, 1]$ *is called a t-conorm iff it fulfills*

1. *For all* $a \in [0, 1]$ *the equation* $\perp(a, 0) = a$ *holds. (0 is the unitary element.)*

2. *For all* $a, b, c \in [0, 1]$ *the equation* $a \leq b \Rightarrow \perp(a, c) \leq \top(b, c)$ *is fulfilled. ($\perp$ is monotonously not decreasing.)*

3. *For all* $a, b \in [0, 1]$ *the equation* $\perp(a, b) = \perp(b, a)$ *holds. (Commutativity)*

4. *For all* $a, b, c \in [0, 1]$ *the equation* $\perp(a, \perp(b, c)) = \perp(\perp(a, b), c)$ *is fulfilled. (Associativity)*

**Remark 1**
1. *Because of the associativity of the t-norm and the t-conorm, we can extend the notation for more than two input fuzzy sets and denote*

   - $\top(a, \top(b, c)) = \top(a, b, c)$
   - $\perp(a, \perp(b, c)) = \perp(a, b, c)$

2. In example 3 we give three pairs of t-norm and t-conorm. When they are used together, the DeMorgan rules

$$\overline{a \bot b} = \overline{a} \top \overline{b} \text{ and } \overline{a \top b} = \overline{a} \bot \overline{b}$$

hold with $\overline{x}$ being the complement of $x$. For a fuzzy set $\mu(x)$, the complement is defined by $\overline{\mu}(x) := 1 - \mu(x)$ [107].

## Example 3
*Some well known pairs of t-norms and t-conorms are to be mentioned:*

1. The usage of $\top_{\min}(a, b) := \min\{a_1, \ldots, a_n\}$ and $\bot_{\max} := \max \{a_1, \ldots, a_n\}$ (see [107]) has the great advantage of the validity of the distributivity in the case with two arguments:

$$\top_{\min}(a, \bot_{\max}(b, c)) = \bot_{\max}(\top_{\min}(a, b), \top_{\min}(a, c)) \qquad \text{and}$$
$$\bot_{\max}(a, \top_{\min}(b, c)) = \top_{\min}(\bot_{\max}(a, b), \bot_{\max}(a, c)).$$

2. The Łukasiewicz-t-norm $\top_{Luka}(a, b) = \max\{0, a+b-1\}$ and the bounded sum $\bot_{Luka}(a, b) = \min\{a + b, 1\}$ are used together. Because of the associativity they can be extended to $n$ arguments and are determined by

$$\top_{Luka}(a_1, \ldots, a_n) = \max\{0, \sum_{i=1}^{n} a_i + 1 - n\} \text{ and}$$
$$\bot_{Luka}(a_1, \ldots, a_n) = \min\{\sum_{i=1}^{n} a_i, 1\}.$$

3. The product-t-norm $\top_{prod}(a, b) = a \cdot b$ and the conorm $\bot_{prod}(a, b) = a + b - a \cdot b$ are another combination that fulfills the DeMorgan rules.

Fuzzy sets enable us to use the graduation, that we are used to in human thinking, to establish classification systems.

## 2.1.2   Fuzzy Classification Systems

There is a wide range of possibilities to use fuzzy sets. We want to restrict our considerations to classification problems. The output in $[0, 1]$ enables us to know whether the classification is clear or the input date belongs to one class nearly as much as to the other.

A fuzzy classification system can be formalized as follows. We have a finite set $\mathcal{R}$ of rules of the form

$$R: \text{ If } x_1 \text{ is } \mu_R^{(1)} \text{ and } \ldots \text{ and } x_m \text{ is } \mu_R^{(m)} \text{ then class is } \mathcal{C}_R,$$

where $\mathcal{C}_R \in \mathcal{C}$ with $\mathcal{C}$ being a finite set of classes. The $\mu_R^{(i)}$ are assumed to be fuzzy sets on the $X_i$, i.e. $\mu_R^{(i)} : X_i \rightarrow [0, 1]$ with $X_i \subset \mathbb{R}$.

**Remark 2**

*Note that choosing a crisp output is not the only possibility. There are several methods to determine a fuzzy output (e.g.[65], [101]). Depending on the purpose of the fuzzy system, a defuzzification method can be chosen. Especially in the applications of fuzzy control, the fuzzy output has to be defuzzified. Examples for defuzzification are choosing the mean of maxima or the center of gravity of the output [54].*

In order to keep the notation simple, we incorporate the fuzzy sets $\mu_R^{(i)}$ directly in the rules. In real systems one would replace them by suitable linguistic values like *positive big, approximately zero*, etc. and associate the linguistic value with the corresponding fuzzy set.

To evaluate the firing degree of a single rule we use the t-norm $\top$, i.e.

$$\mu_R(p_1, \ldots, p_m) = \top_{i \in \{1, \ldots, m\}} \left\{ \mu_R^{(i)}(p_i) \right\}. \tag{2.1}$$

The degree to which the point $(p_1, \ldots, p_m)$ is assigned to class $\mathcal{C}$, is determined by

$$\mu_{\mathcal{C}}^{(\mathcal{R})}(p_1, \ldots, p_m) = \max \left\{ \mu_R(p_1, \ldots, p_m) \mid \mathcal{C}_R = \mathcal{C} \right\}, \tag{2.2}$$

i.e. by the maximum-t-norm.

Finally, we have to assign the point $(p_1, \ldots, p_m)$ to a unique class (defuzzification) by

$$\mathcal{R}(p_1, \ldots, p_m) = \begin{cases} \mathcal{C}_i & \text{if } \mu_{\mathcal{C}_i}^{(\mathcal{R})}(p_1, \ldots, p_m) > \mu_{\mathcal{C}_j}^{(\mathcal{R})}(p_1, \ldots, p_m) \\ & \text{for all } \mathcal{C}_j \in \mathcal{C}, j \neq i \\ unknown & \text{otherwise} \end{cases}$$

This means that we assign the point $(p_1, \ldots, p_m)$ to the class of the rule with the maximum firing degree. Note that we denote by $\mathcal{R}$ the set of rules as well as the associated classification mapping based on these classification rules.

**Example 4**

*Let us assume that we have two rules*

$R_1$ : *If $x_1$ is approximately 1 and $x_2$ is approximately 2, then class is $\mathcal{C}_1$*
$R_2$ : *If $x_1$ is approximately 2 and $x_2$ is approximately 1, then class is $\mathcal{C}_2$*

*with the fuzzy sets as depicted in figure 2.3. We use the min-t-norm. With*

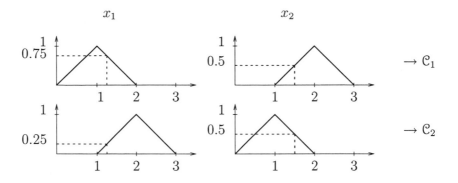

Figure 2.3: The rulebase for example 4

the input value $x = (1.25, 1.5)$, the membership degrees of $x$ to the fuzzy sets
are

$$\mu_{R_1}(x) = \min\{\mu_{R_1}^{(1)}(1.25), \mu_{R_1}^{(2)}(1.5)\} = \min\{0.75, 0.5\} = 0.5$$
$$\mu_{R_2}(x) = \min\{\mu_{R_2}^{(1)}(1.25), \mu_{R_2}^{(2)}(1.5)\} = \min\{0.25, 0.5\} = 0.25$$

Therefore the point $x$ can be assigned to class $\mathcal{C}_1$.

**Example 5**
At airports, the arrival times and delays of arriving aircrafts depend on many
factors. One amongst them is weather. We will explain this more detailed in
chapter 10. Here we only want to demonstrate how rules can look like, that
describe the dependency of the arrival time on the weather conditions.

$R_1$ :   If temperature higher than 22.5 degrees,
            clouds higher than 4500 ft and
            south wind component lower than 9.2 KT        then flight is in time.
$R_2$ :   If temperature lower than 20 degrees,
            clouds higher than 4500 ft and
            south wind component greater than 9.2 KT,   then flight is delayed.

Figure 2.4 demonstrate how these rules could look like. The outcome can
be considered as a possibility how the arrival time of aircrafts will be (see
[22, 102, 89]).
The example is simplified. More detailed information can be found in [87].

An early overview on fuzzy classification is given in [68].

Temperature  Heigth of Clouds   Southwind component

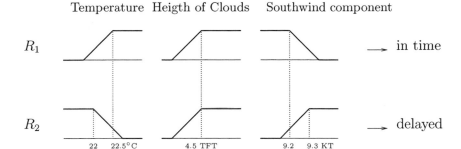

Figure 2.4: The rulebase for example 5

Figure 2.5: The Minimum-t-norm

## 2.2  Examples for Different t-Norms

Nürnberger et al. [79, 80] investigated the class boundaries of two- and three-dimensional data that can be generated by fuzzy classification systems with different t-norms.

The separation that is given by a fuzzy classification system strongly depends on the t-norm that is used. Compare e.g the minimum-t-norm and the product-t-norm as shown in figure 2.5 and 2.6.

These examples show the two-dimensional case. The fuzzy sets can be found on the edges of the graphics. The third dimension demonstrated by the contour lines shows the membership degree of the rule for the two-dimensional input values.

Although the separation looks the same on the right side where the fuzzy sets are regularly distributed, the separation is quite different when only two rules are to be considered as shown on the left side.

In higher dimensions the differences between the t-norms become even more relevant.

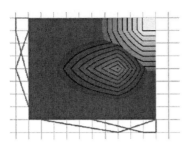 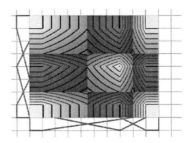

Figure 2.6: The Product-t-norm

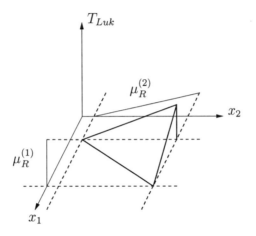

Figure 2.7: The Łukasiewicz-t-norm

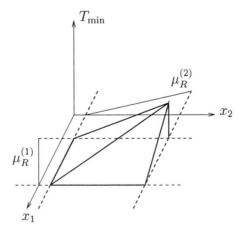

Figure 2.8: The Minimum-t-norm

When comparing the Łukasiewicz-t-norm (figure 2.7) and the min-t-norm (figure 2.8), we realize, that rules based on the min-t-norm tend to fire in a wider section of the space than those based on the Łukasiewicz-t-norm. In chapter 3, we will examine the min-t-norm and point out its limitations, and in chapter 7, we will demonstrate how to use the advantages of the Łukasiewicz-t-norm (for further differences see e.g. [26]).

## 2.3 Advantages of Fuzzy Classifiers

When using linguistic variables as "small" or "fast" in the rules, we have a classification system that is well interpretable, as the rules can intuitively be understood. No translation of the output is necessary.

Expert knowledge can be represented by such a system. A human expert usually is able to give rules with linguistic variables, but rarely with concrete values. Therefore we need to transform this knowledge. If we are able to form fuzzy sets from the linguistic terms, then we can also construct the rules.

When choosing the right t-norm, we will see in chapter 5 that these rules represent separating hyperplanes, that divide the data space into sections, and these sections can be assigned to the classes. This correlation leads us to a method to interpret the rules by using their geometric visualization.

But first we will examine the max-min-classifier. The following chapter will illustrate and prove formally the limitations.

# Chapter 3

# Fuzzy Max-Min Classifiers

In this chapter we consider fuzzy classification systems using the max-min inference scheme to classify an unknown datum on the basis of maximum matching, i.e. assigning it to the class appearing in the consequent of the rule whose premise fits best. We basically show that this inference scheme locally takes only two attributes (variables) into account for the classification decision, i.e. that fuzzy max-min classifiers decide locally on the basis of two attributes.

A theoretical analysis of fuzzy classification systems is presented in [44, 46]. It was demonstrated that often only approximate solutions can be constructed when crisp sets are used. Figure 3.1 shows an approximate solution for a linearly separable problem, i.e. a separation that can be described by a linear threshold function. An exact solution is not possible, because the separation is always parallel to the coordinates.

In the case of two-dimensional data, classification problems can exactly be solved, when the classes can be separated by piecewise monotone functions.

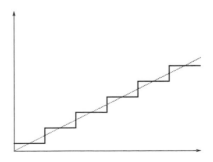

Figure 3.1: Crisps sets approximate a linear separation

For higher dimensions, linearly separable problems can only be solved, if the separation hyperspace is always locally parallel to two coordinates, because max-min-classifiers decide locally on the basis of two attributes.

When the Łukasiewicz t-norm is allowed instead of the minimum or the maximum is replaced by the bounded sum, arbitrary linearly separable classification problems can be solved by fuzzy classification systems, i.e. problems where the classes are separated by a (hyper-)plane. However, fuzzy max-min classification systems cannot solve arbitrary linearly separable classification problems for data with more than two attributes. If the separating hyper-plane depends on more than two variables, fuzzy max-min classification systems can only provide an approximate solution for the classification problem.

In the following sections we generalize this result and show that in principle fuzzy max-min classification systems determine the class locally on the basis of only two attributes. The chapter is organized as follows. The next section briefly reviews the structure of fuzzy max-min classification systems. Then we introduce the necessary basic definitions, that we need, and present our main theorem in section 3.2. Section 3.3 contains he construction that proves the main theorem. Some technical requirements needed in the prove can be found in the appendix A.

## 3.1   Fuzzy Max-Min Classification Systems

We consider the following classification problem. We have a finite number of classes $\mathcal{C}_1, \ldots, \mathcal{C}_c$. Each class represents a subset of the space $\mathbb{R}^m$ or the unit cube $[0, 1]^m$. Therefore, we identify each class with its corresponding subset. We assume that the classes are pairwise disjoint, but we do not require that they cover the whole space, i.e. there might be data that are unclassified.

In practical applications, the situation is usually as follows: A finite set of data including the classes to which the data belong is given. The problem is to find a classifier that – in the best case – assigns all the given (training) data to the corresponding classes and extends the classification also to unknown data in a reasonable way. We do not discuss here, how such a classifier can be learned from data. This can be found e.g. in [58].

We are interested in how flexible a fuzzy classifier can be. Therefore, we assume that the corresponding classes are already known for all possible data, and examine, whether this classification problem can be solved by a fuzzy classifier.

First, we restrict our investigations to classification problems with only two classes $\mathcal{C}^+$ and $\mathcal{C}^-$. However, our results can easily be extended to classification problems with more than two classes. In order to see whether the class $\mathcal{C}_1$ can be separated correctly from the other classes $\mathcal{C}_2, \ldots, \mathcal{C}_c$, we simply have to combine the classes $\mathcal{C}_2, \ldots, \mathcal{C}_c$ to one new class and again have a classification with only two classes.

The rules are defined as described in the precedent chapter and evaluated by interpreting the conjunction in terms of the minimum, i.e. $\top = \top_{min}$, and we defuzzify by assigning the point $(p_1, \ldots, p_m)$ to the class with the maximum firing degree.

When we assume that the fuzzy sets appearing in the rules are continuous, then $\mathcal{C}^+$ and $\mathcal{C}^-$ are open sets. This means that when a point $(p_1, \ldots, p_m)$ is assigned to the class $\mathcal{C}^+$, then there is a neighborhood of this point in which all points are also assigned to $\mathcal{C}^+$. The same holds for the class $\mathcal{C}^-$. We are interested in the class boundaries, i.e. the set of points that are classified as *unknown*. A typical situation for a point classified as *unknown* is the following: There is exactly one rule firing with the maximum degree for class $\mathcal{C}^+$ and also exactly one rule firing with the maximum degree for class $\mathcal{C}^-$. For each of these rules the firing degree is determined by just one attribute for which the membership degree to the corresponding fuzzy set in the rule yields the firing degree (the minimum of the t-norm). Let us for the moment assume that $x_1$ is the corresponding attribute for class $\mathcal{C}^+$ and $x_2$ for class $\mathcal{C}^-$. This means that the firing degree for class $\mathcal{C}^+$ and $\mathcal{C}^-$ does not change, when any of the values of the attributes $x_3, \ldots, x_m$ slightly changes. In this sense, the classification *depends* in this situation *locally* only *on the two attributes $x_1$ and $x_2$.*

However, the above considerations are only correct in this special case where the maximum firing degree for each class is determined by just one rule and the minimum in the corresponding rules is determined by just one variable each.

When there are more than just two rules firing with maximum degree, the situation is different. Let us consider the six rules

$R_1$: If $x_1$ is positive small and $x_2$ is anything and $x_3$ is anything
   then class is $\mathcal{C}^+$

$R_2$: If $x_1$ is anything $x_2$ is positive small and $x_3$ is anything
   then class is $\mathcal{C}^+$

$R_3$: If $x_1$ is anything and $x_2$ is anything and $x_3$ is positive small
   then class is $\mathcal{C}^+$

$R_4$: If $x_1$ is negative small and $x_2$ is anything and $x_3$ is anything
   then class is $\mathcal{C}^-$,

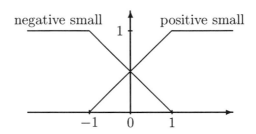

Figure 3.2: Two fuzzy sets

$R_5$: If $x_1$ is anything and $x_2$ is negative small and $x_3$ is anything
     then class is $\mathcal{C}^-$,

$R_6$: If $x_1$ is anything and $x_2$ is anything and $x_3$ is negative small
     then class is $\mathcal{C}^-$,

where the fuzzy sets *positive small* and *negative small* are chosen as illustrated in figure 3.2 and the fuzzy set corresponding to *anything* is the constant function 1.
When we consider the point $(0,0,0)$, we have

$$\mu_{\mathcal{C}+}^{(\mathcal{R})}(0,0,0) \;=\; 0.5$$

and

$$\mu_{\mathcal{C}-}^{(\mathcal{R})}(0,0,0) \;=\; 0.5,$$

i.e. $(0,0,0)$ is classified as unknown. But when we increase any of the three variables $x_1$, $x_2$, or $x_3$, the resulting point is classified to $\mathcal{C}^+$ and when we decrease any of these three variables the resulting point is classified to $\mathcal{C}^-$. This means that the classification near $(0,0,0)$ depends on all three attributes. If we choose $\varepsilon > 0$, we have $(\varepsilon,0,0),(0,\varepsilon,0),(0,0,\varepsilon) \in \mathcal{C}^+$ and $(-\varepsilon,0,0),(0,-\varepsilon,0),(0,0,-\varepsilon) \in \mathcal{C}^-$. Therefore, we cannot say that fuzzy max-min classification systems generally decide locally on the basis of two variables. However, the above described example can be seen as an exception and we can show the following: When there is a point on the class boundary where the classification depends (locally) on more than two variables, then in any neighborhood of this point there is another point on the class boundary where the classification depends locally only on at most two variables. In this sense, a boundary point where the classification depends on more than two variables can be seen as some kind of singularity.

So far we have not made any assumptions on the fuzzy sets. We require that they are continuous and that they have a local one-sided Taylor expansion everywhere: If $\mu$ is a fuzzy set, then for any $x_0 \in \mathbb{R}$ there is $\varepsilon > 0$ and there exist power series

$$\sum_{k=0}^{\infty} a_k^{(l)}(x - x_0)^k \quad \text{and} \quad \sum_{k=0}^{\infty} a_k^{(r)}(x - x_0)^k,$$

so that

$$\mu(x_0 - h) = \sum_{k=0}^{\infty} a_k^{(l)} h^k$$

and

$$\mu(x_0 + h) = \sum_{k=0}^{\infty} a_k^{(r)} h^k$$

hold for all $0 < h < \varepsilon$.

Note that membership functions typically used in application like piecewise linear functions (for instance triangular or trapezoidal fuzzy sets) or Gaussian membership functions fulfill this property. In the following section we will see, why we need this technical condition.

## 3.2 Basic Definitions

We consider a fuzzy max-min classification system as it was described in the previous section.

**Definition 4**
*The set $\mathcal{D}$ of the points that have the same membership degree to $\mathcal{C}^+$ as to $\mathcal{C}^-$ is called separating set. A point $P \in \mathcal{D}$ is called a separating point.*

$\delta^{(i)}$ denotes the vector that has $\delta$ as the $i^{th}$ component and 0 for the other components.

As we have already mentioned in the previous section, we require that the fuzzy sets have a local one-sided Taylor expansion. In order to illustrate what can happen, if we refrain from this restriction, we consider the following example.

**Example 6**
*We consider a fuzzy classification system for data with just one attribute $x$ with the following two rules:*

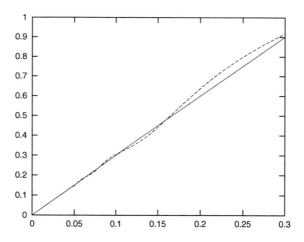

Figure 3.3: Two fuzzy sets

$R_1$: If $x$ is $\mu_1$ then class is $\mathcal{C}^+$
$R_2$: If $x$ is $\mu_2$ then class is $\mathcal{C}^-$

where the fuzzy sets $\mu_1$ and $\mu_2$ are defined by

$$\mu_1(x) = \begin{cases} 0 & \text{if } x < 0 \\ 3x & \text{if } 0 \le x \le \frac{1}{3} \\ 1 & \text{otherwise} \end{cases}$$

and

$$\mu_2(x) = \begin{cases} 0 & \text{if } x \le 0 \\ 3x - x^2 \cdot \sin(\frac{1}{x}) & \text{if } 0 < x \le \frac{1}{3} \\ 1 - \frac{1}{9} \cdot \sin(3) & \text{otherwise.} \end{cases}$$

Figure 3.3 illustrates these two fuzzy sets.
Note that $\mu_2$ is continuous, even differentiable on $]-\infty, \frac{1}{3}[$, and monotonous
(the first derivative is positive), but does not have a local one-sided Taylor
expansion at $x_0 = 0$. By applying L'Hospital's rule, it can be proved easily
that $\mu_2$ is continuous and the first derivative exists on $]-\infty, \frac{1}{3}[$.
The point 0 is a separating point for this fuzzy classification system. But for
any interval $[0, \varepsilon]$, no matter how small we choose $\varepsilon > 0$, there are points in
this interval that are classified to $\mathcal{C}^+$ and also points that are classified to
$\mathcal{C}^-$.

By requiring that each fuzzy set is continuous and has a local one-sided
Taylor expansion everywhere, we can not have such a strange situation as in
the above described example. When we are in a separating point and consider

one variable that we want to change in one direction by a very small value, then we can say that we always end up in the same class (or always remain in the separating set), as long as the change is small enough. An 'oscillation' between the classes as in the example is not possible. The following shows that our fuzzy classification systems have this property.

**Lemma 1**

*For each point p and for each coordinate $p_i$ of p, we have*

$$(\exists\, A, B \in \{\mathcal{C}^+, \mathcal{C}^-, \mathcal{D}\})(\exists\, \varepsilon > 0)\ (\forall\, 0 < \delta < \varepsilon)(p + \delta^{(i)} \in A \,\wedge\, p - \delta^{(i)} \in B). \tag{3.1}$$

**Proof:** If $p$ is not a separating point, then $p$ belongs to $\mathcal{C}^+$ or $\mathcal{C}^-$. Since these sets are open, a sufficiently small variation of any variable will not lead out of these sets. Therefore, we only need to consider separating points.

Let us assume, that we want to increase the variable $p_i$. It is easy to determine, how a (sufficiently) small increase of $p_i$ will alter the firing degree $\mu_{\mathcal{C}^+}^{(\mathcal{R})}(p)$ for class $\mathcal{C}^+$, if equality will not lead to remaining in $\mathcal{D}$. When the change of $p_i$ influences the firing degree $\mu_{\mathcal{C}^+}^{(\mathcal{R})}(p)$ at all, we only need to consider the rules firing for class $\mathcal{C}^+$ in which $p_i$ determines the minimum. It is now easy to determine which will take over when we increase $p_i$: We need to know which fuzzy set for $p_i$ will yield the strongest change. Since we can compute the Taylor expansions of each fuzzy set for $p_i$, we can easily solve this problem using the corollaries 10 and 11 in appendix A. We can do the same for the rules for class $\mathcal{C}^-$. Finally, we have to decide for which class we have the strongest change. But this can be done again by making use of lemma 9. □

Note that we only need the Taylor expansions for the proof of lemma 1. The essential property that we are interested in is (3.1).

**Definition 5**

*Let p be a point of the separating set $\mathcal{D}$. p is called a proper separating point, when*

$$(\forall\, \varepsilon > 0)(\,\exists\, p', p'' \in \mathcal{N}_\varepsilon(p)) : (p' \in \mathcal{C}^+ \text{ and } p'' \in \mathcal{C}^-)$$

*holds, where $\mathcal{N}_\varepsilon(p)$ denotes the $\varepsilon$-neighborhood of p.*

This means that for $\mathcal{C}^-$ as well as for $\mathcal{C}^+$ there exists a direction in which the set can be reached in an arbitrarily small distance from $p$.

**Definition 6**

*Let p be a proper separating point and $x_i$ a single variable with the value $p_i$ for p.*

1. $x_i$ is called relevant at $p$ iff

$$(\exists\, \varepsilon > 0)(\forall\, 0 < \delta^{(i)} < \varepsilon): \quad \{p + \delta^{(i)}, p - \delta^{(i)}\} \cap \mathcal{C}^+ \neq \emptyset \text{ and}$$
$$\{p + \delta^{(i)}, p - \delta^{(i)}\} \cap \mathcal{C}^- \neq \emptyset.$$

This means that increasing $p_i$ leads into one set and decreasing $p_i$ leads into the other one.

2. $x_i$ is called semi-relevant (for $\mathcal{C}^+$) at $p$ iff

$$(\exists\, \varepsilon > 0)(\forall\, 0 < \delta^{(i)} < \varepsilon): \quad \{p + \delta^{(i)}, p - \delta^{(i)}\} \subset \mathcal{C}^+ \cup \mathcal{D} \text{ and}$$
$$\{p + \delta^{(i)}, p - \delta^{(i)}\} \cap \mathcal{C}^+ \neq \emptyset$$

This means that we can reach only $\mathcal{C}^+$ and not $\mathcal{C}^-$ by varying $p_i$ by an arbitrarily small distance. In the same way we define the notion "semi-relevant for $\mathcal{C}^-$".

3. $x_i$ is called irrelevant at $p$ iff

$$(\exists\, \varepsilon > 0)(\forall\, 0 < \delta^{(i)} < \varepsilon) : \{p + \delta^{(i)}; p - \delta^{(i)}\} \subseteq \mathcal{D}.$$

This means that it is impossible to reach either $\mathcal{C}^+$ or $\mathcal{C}^-$ by varying $p_i$ by an arbitrarily small amount.

Lemma 1 guarantees that a proper separating point is either relevant, semi-relevant or irrelevant.

**Theorem 2**
Let $p$ be a proper separating point and $\varepsilon > 0$. Then there exists a proper separating point $p'$ in the neighborhood $\mathcal{N}_\varepsilon(p)$ of $p$ that has at most two variables that are relevant or semi-relevant.

The proof of this theorem will be given in the next section, where we actually show constructively, how to obtain the point $p' \in \mathcal{N}_\varepsilon(p)$ that has only two relevant or semi-relevant variables.

**Remark 3**
There is no reason in considering a point $p$ that does not belong to $\mathcal{D}$. Because of $\mathcal{C}^+$ and $\mathcal{C}^-$ being open sets, there is always a neighborhood of $p$ that is completely contained in $\mathcal{C}^+$, respectively $\mathcal{C}^-$. In this sense, points that are in one of the classes, do not have relevant variables.
In case of $p$ being an inner point of $\mathcal{D}$ we can use the same argument.

## 3.3 A Point with only Two Relevant Variables

We now provide the proof of theorem 2. Without loss of generality we assume that $x_1, \cdots, x_n$ are the relevant and semi-relevant variables and $x_{n+1}, \cdots, x_m$ are the irrelevant ones.

**Remark 4**
*We consider a point $p = (p_1, \cdots, p_m) \in \mathbb{R}^m$. The set of rules that are firing in $p$ with the maximum degree $\mu_{\mathcal{C}+}^{(\mathcal{R})}(p) = \mu_{\mathcal{C}-}^{(\mathcal{R})}(p)$ is denoted by $\mathcal{R}_p$.*

1. *When we vary an attribute of $p$, the variation has to be sufficiently small, so that for the new point $p' = (p_1, \cdots, p_i', \cdots, p_m)$ there is no new rule $R$ with $R \in \mathcal{R}_{p'}$ but $R \notin \mathcal{R}_p$.*

2. *When we consider two fuzzy-sets $\mu = \mu_R^{(i)}$ and $\nu = \nu_{R'}^{(i)}$ for one variable $x = x_i$, we want to know which one is increasing faster, when we vary $x$. Since the Taylor expansions at $x_0$ in the considered direction of the two functions exist, we can compute the (directed) derivatives at $x_0$. When the first $n$ derivatives are equal, but for the $(n+1)^{th}$ derivative we have $\mu^{(n+1)}(x_0) > \nu^{(n+1)}(x_0)$, then within $\mathcal{N}_\varepsilon(x_0)$ $\mu$ is increasing faster than $\nu$, when increasing $x$, and the other way round, when decreasing $x$.*

   *For this comparison we only need the coefficients of the Taylor expansion up to that term that is different for $\mu$ and $\nu$. We will explain this more detailed in the appendix A.*

3. *The variation $\varepsilon$ must also be sufficiently small, so that $\mu$ and $\nu$ do not 'overtake' each other. This means the following: When $\mu^{(n+1)}(x_0) > \nu^{(n+1)}(x_0)$, then $\mu(x) > \nu(x)$ for all $x > x_0$, $x \in \mathcal{N}_\varepsilon(x_0)$, and $\mu(x) < \nu(x)$ for all $x < x_0$, $x \in \mathcal{N}_\varepsilon(x_0)$.*

Because of part 1 of remark 4 the irrelevant variables stay irrelevant, so that we do not need to consider them at all. For the proof of theorem 2, we only need to consider the case that there are at least three relevant or semi-relevant variables at the point $p$.

### 3.3.1 First case: One Dominating Variable for Each Rule

**Definition 7**
*Let $R$ be a rule and let $x_i$ be a variable. $x_i$ is called dominating at point $p$ iff*

$$\mu_R(p) = \mu_i^{(R)}(p_i)$$

with $\mu_i^{(R)}(p_i)$ being the membership degree of the fuzzy set for rule $R$ at the the $i$th variable for point $p = (p_1, \ldots, p_m)$.

The first case to be considered is the most simple one, where each rule of $\mathcal{R}_p$ has only one dominating variable at point $p$, i.e. every rule contributing to the maximum firing degree has just one dominating variable.

Changing one variable $x_i$ and considering a single rule (without loss of generality firing for $\mathcal{C}^+$), we have the following: When $\mu_i$ is increasing with the change of $x_i$, then the rule leads to a decision for $\mathcal{C}^+$; the case of $\mu_i$ remaining constant is trivial, and when $\mu_x$ is decreasing, the rule is not relevant any more for the classification of $p$.

## Construction of $\Lambda_i^+$ and $\Lambda_i^-$

$\mathcal{R}_p^{(i)}$ denotes the set of rules firing in $p$ with the dominating variable $x_i$, i.e. the rules $R \in \mathcal{R}_p$ with $\mu_R^{(i)}(p_i) = \mu_R(p)$. For our purpose, it is possible to combine the firing degrees given by the rules of $\mathcal{R}_p^{(i)}$ firing for $\mathcal{C}^+$ into one function $\Lambda_i^+$.

$\Lambda_i^+$ describes the membership degree for the class $\mathcal{C}^+$, when we vary the attribute $p_i$ of $p$ and consider only the rules of $\mathcal{R}_p^{(i)}$:

$$\Lambda_i^+(\delta) := \mu_{\mathcal{C}^+}^{(\mathcal{R}_p^{(i)})}(p + \delta^{(i)}) = \max\{\mu_R^{(i)}(p + \delta^{(i)}) \mid R \in \mathcal{R}_p^{(i)}, \ C_R = \mathcal{C}^+\}$$

The $\Lambda_i^-$ are defined analogously. When there are different rules firing with maximum degree at $p$ that have all $x_i$ as the only dominating variable, then we normally get a sharp bend at $\delta = 0$ (see figure 3.4).

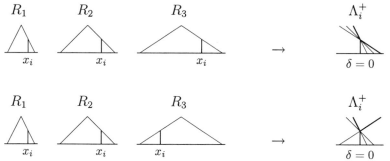

Figure 3.4: Two examples for the construction of $\Lambda_i^+$

When all rules firing at $p$ for $\mathcal{C}^+$ with the dominating variable $x_i$ are combined into $\Lambda_i^+$, then we can consider $\Lambda_i^+$ as the only function giving the degree to which $p$ belongs to $\mathcal{C}^+$ with respect to the dominating variable $x_i$.

$\Lambda_i^-$ is constructed in the same way, and if e.g. for $\mathcal{C}^+$ and $x_i$ there is no such rule, then we have $\Lambda_i^+ \equiv 0$.

When each attribute $p_i$ is changed by $\delta_i$, then we denote the vector incorporating all changes by $\bar{\delta} := \sum_{i=1}^n \delta_i^{(i)}$. The functions $\Lambda_M^+$ and $\Lambda_M^-$ of the total membership degree of $p' := p + \bar{\delta}$ to $\mathcal{C}^+$ and $\mathcal{C}^-$ are calculated in the following way:

$$\Lambda_M^+(p') = \Lambda_M^+(p + \bar{\delta}) := \max\{\Lambda_i^+(\delta_i) \mid i \in \{1, \cdots, n\}\},$$

and analogously for $\Lambda_M^-$.

**Moving towards a point with only two relevant variables**

We consider $\Lambda_i^+$ and $\Lambda_i^-$ instead of the individual rules.

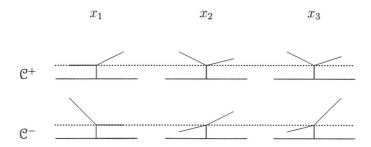

Figure 3.5: An example with three variables for $\Lambda_i^+$ and $\Lambda_i^-$

For the procedure of finding $p' \in \mathcal{N}_\varepsilon(p) \cap \mathcal{D}$ with $p'$ having just two relevant variables we choose a variable $x_i$ that is relevant or semi-relevant for $\mathcal{C}^+$ and another variable $x_j$ that is relevant or semi-relevant for $\mathcal{C}^-$. We take $p_i$ and vary it by the distance $\delta_i \neq 0$, so that $\Lambda_i^+(\delta_i) > \Lambda_i^-(\delta_i)$ and $\Lambda_i^+(\delta_i) > \Lambda_k^+(0) = \Lambda_k^-(0)$ for $k \neq i$.
We obtain a point $p'' = p + \delta_i^{(i)}$ that belongs to $\mathcal{C}^+$, but we are looking for a point $p' \in \mathcal{D}$. Now we can alter the attribute $p_j$ of $p$ by $\delta_j \neq 0$ towards that direction where $\Lambda_j^-$ grows faster than $\Lambda_j^+$, so that $\Lambda_j^-(\delta_j) > \Lambda_j^+(\delta_j)$ holds. We choose $\delta_j$ in such a way that after this step we have $\Lambda_j^-(\delta_j) = \Lambda_i^+(\delta_i) > \Lambda_k^+(0) = \Lambda_k^-(0)$ for $i \neq k \neq j$. Then $p' = p + \delta_i^{(i)} + \delta_j^{(j)}$ is an element of $\mathcal{D}$. Now the point is reached, where the variables $x_k$ for $i \neq k \neq j$ are not dominating anymore in any rule. Therefore they are irrelevant. Only $x_i$ and $x_j$ are relevant (not semi-relevant) variables in $p'$. For the total membership

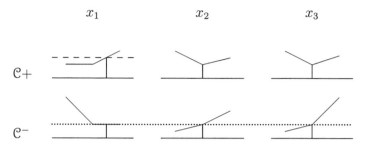

Figure 3.6: The same example after pushing $x_1$

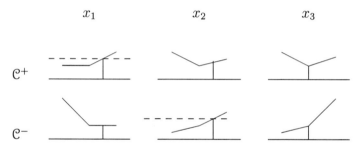

Figure 3.7: The example after changing $x_2$

degrees $\Lambda_M^+$ and $\Lambda_M^-$ of the point $p'$ to $\mathcal{C}^+$ and $\mathcal{C}^-$, we obtain

$$\Lambda_M^+(p + \bar{\delta}) = \max\{\Lambda_i^+(\delta_i), \Lambda_j^+(\delta_j)\} \text{ and } \Lambda_M^-(p + \bar{\delta}) = \max\{\Lambda_i^-(\delta_i), \Lambda_j^-(\delta_j)\}.$$

Because of having changed the variables sufficiently small, $\mathcal{R}_p'$ does not contain any new rules. Therefore, we have $\Lambda_{x_i+\delta_i}^+(\varepsilon) = \Lambda_{x_i}^+(\varepsilon + \delta_i)$ in $\mathcal{N}_\varepsilon(p)$. This means that within the domain we can still use the same functions of membership degree for $p'$ as for $p$.

When taking $p_i' = p_i + \delta_i$ and changing it into that direction where the membership degree for $\mathcal{C}^+$ is increasing, this yields a decision for $\mathcal{C}^+$. When moving $p_i'$ towards the other direction the membership degree for $\mathcal{C}^+$ is lowered and we get a decision for $\mathcal{C}^-$. The same applies to $p_j' = p_j + \delta_j$ and $\mathcal{C}^-$, so that $x_i$ and $x_j$ are both relevant variables.

**Remark 5**
*Because the fuzzy sets have (directed) derivatives, it is possible to vary $p_i$ by an arbitrary small value, so that the value needed to vary $p_j$ is small enough so that $p'$ is in the $\varepsilon$-neighborhood $\mathcal{N}_\varepsilon(p)$ of $p$.*

**Proof:** $\mu_i$ and $\mu_j$ have a bounded slope on $\mathcal{N}_\varepsilon(p)$, because they are also differentiable on the closure $\overline{\mathcal{N}}_\varepsilon(p)$. Therefore, it is possible to calculate $\max_{p_i' \in \overline{\mathcal{N}}_{\frac{\varepsilon}{n}}(p_i)} \{||\mu_i(p_i') - \mu_i(p_i)||\}$ and $\max_{p_j' \in \overline{\mathcal{N}}_{\frac{\varepsilon}{n}}(p_j)} \{||\mu_j(p_j') - \mu_j(p_j)||\}$. We define

$$\Delta\mu := \min\{ \max_{p_i' \in \mathcal{N}_{\frac{\varepsilon}{n}}(p_i)} ||\mu_i(p_i') - \mu_i(p_i)||; \max_{p_j' \in \mathcal{N}_{\frac{\varepsilon}{n}}(p_j)} ||\mu_j(p_j') - \mu_j(p_j)||\}.$$

Now we can choose $p_i' \in \mathcal{N}_{\frac{\varepsilon}{n}}(p_i)$, so that $||\mu_i(p_i') - \mu_i(p_i)|| = \Delta\mu$, and $p_j' \in \mathcal{N}_{\frac{\varepsilon}{n}}(p_j)$ so that $||\mu_j(p_j') - \mu_j(p_j)|| = \Delta\mu$.

When taking the Euclidian norm, we obtain

$$\begin{aligned} ||p' - p|| &= \sqrt{\sum_{k=1}^n (p_k' - p_k)^2} \\ &\leq \sqrt{\sum_{k=1}^n \left(\frac{\varepsilon}{n}\right)^2} = \sqrt{n \cdot \left(\frac{\varepsilon}{n}\right)^2} \\ &= \frac{1}{\sqrt{n}}\varepsilon < \varepsilon, \end{aligned}$$

which proves that $p' \in \mathcal{N}_\varepsilon(p)$ holds. When considering another norm we just have to take $\frac{\varepsilon}{\alpha}$ with (another $\alpha$ instead of $n$) instead of $\frac{\varepsilon}{n}$. $\qquad\square$

## 3.3.2 Second case: Rules with more than One Dominating Variable

In this section we consider the case that there are rules contributing to the maximum firing degree with more than one dominating variable. If there are also rules having only one relevant variable, we use $\Lambda_i^+$ and $\Lambda_i^-$ as already described for these rules.

First we consider a single rule $R$ with two or more dominating variables. When the dominating variable $x_i$ is varied there are three possibilities: If $\mu_i$ is increasing, $x_i$ is not dominating any more, so that there is one dominating variable less in this rule, but the rule is still firing. The case of $\mu_i$ remaining constant is trivial, and if $\mu_i$ is decreasing, the firing degree of the whole rule is decreasing. In no case the firing degree $\mu_R$ of the rule $R$ is increasing.

Because of $x_i$ being a relevant or semi-relevant variable, varying $p_i$ has to lead into $\mathcal{C}^+$ or $\mathcal{C}^-$. Without loss of generality we consider $\mathcal{C}^+$. We can reach $\mathcal{C}^+$ iff

1. $\Lambda_i^+$ is increasing, and if we have that $\Lambda_i^- \not\equiv 0$ is also increasing, then $\Lambda_i^+$ has to increase faster than $\Lambda_i^-$.

2. Every rule $R$ firing for $\mathcal{C}^-$ with maximum degree has $x_i$ as a dominating variable, and in every rule $\mu_R^{(x_i)}$ is decreasing, so that the total membership degree $\mu_{\mathcal{C}^-}^{(\mathcal{R})}$ for $\mathcal{C}^-$ is decreasing. If all the rules firing for

$\mathcal{C}^+$ with maximum degree have $x_i$ as a dominating variable, too, then $\mu_{\mathcal{C}^-}^{(\mathcal{R})}$ has to decrease faster than $\mu_{\mathcal{C}^+}^{(\mathcal{R})}$.

In any case there are at least two variables $x_i$ and $x_j$ with (1.) being satisfied for $x_i$ for reaching $\mathcal{C}^+$ and for $x_j$ for reaching $\mathcal{C}^-$ or with (2.) being satisfied for $x_i$ and $\mathcal{C}^+$ and for $x_j$ and $\mathcal{C}^-$, because of the following:
Suppose that the variables $x_k$, $k \in A \subseteq \{1, \cdots, n\}$, being relevant or semi-relevant for (without loss of generality) $\mathcal{C}^+$ are satisfying (1.), when $p_k$ is moved by $\delta_k$. This means that $\Lambda_i^+ \not\equiv 0$ holds for these variables in the direction of the movement. Furthermore, suppose that the variables $x_l$, $l \in B \subseteq \{1, \cdots, n\}$, being relevant or semi-relevant for $\mathcal{C}^-$ are satisfying (2.), when $p_l$ is moved by $\delta_l$. This means that all these variables are dominating in every rule firing for $\mathcal{C}^+$. If we have more than two relevant or semi-relevant variables, this is a contradiction, because there is at least one rule for $\mathcal{C}^+$ having only one dominating variable $x_i$, so that the other variables $x_l$, $l \neq i$, cannot satisfy (2).
Now we have to consider the two cases that are left:

1. At least for two variables (without loss of generality $x_i$ and $x_j$) there are rules with only this variable dominating and with $\mu_{\mathcal{C}^+}$ increasing when varying $p_i$ and $\mu_{\mathcal{C}^-}$ increasing when varying $p_j$. Then the procedure is the same as described in section 3.3.1.

2. Without loss of generality every rule firing for $\mathcal{C}^-$ has $x_i$ as one dominating variable, and in each such rule $\mu_R^{(i)}$ is decreasing when $p_i$ is varied into the right direction. The same is the case for $x_j$ and $\mathcal{C}^+$.

   We vary $p_i$ by $\delta_i \neq 0$, so that this leads into $\mathcal{C}^+$, because $\mu_R^{(i)}$ is decreasing in every rule $R$ firing for $\mathcal{C}^-$ with maximum degree and with it $\mu_{\mathcal{C}^-}^{(\mathcal{R})}$.

   When there is at least one rule $R$ firing for $\mathcal{C}^+$ that does not have $x_i$ as a dominating variable or with $\mu_R^{(i)}$ not decreasing, then $\mu_R^{\mathcal{C}^+}$ is not changing. Otherwise $\mu_R^{(i)}$ does decrease more slowly, so that we still have $\mu_{\mathcal{C}^+}^{(\mathcal{R})}(x + \delta_i^{(i)}) > \mu_{\mathcal{C}^-}^{(\mathcal{R})}(x + \delta_i^{(i)})$, because $x_i$ is relevant or semi-relevant for $\mathcal{C}^+$.

   After this the variation of $p_j$ by $\delta_j$ leads back to $\mathcal{D}$, because for $\mathcal{C}^+$ every membership degree $\mu_R^{(j)}$ is decreasing until

$$\mu_{\mathcal{C}^-}^{(\mathcal{R})}(p + \delta_i^{(i)} + \delta_j^{(j)}) = \mu_{\mathcal{C}^+}^{(\mathcal{R})}(p + \delta_i^{(i)} + \delta_j^{(j)}).$$

This completes the proof of theorem 2.

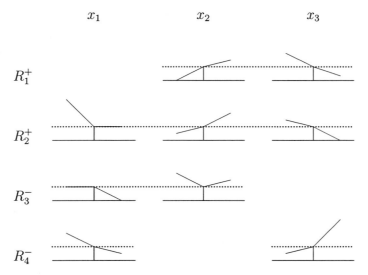

Figure 3.8: An example for three semi-relevant variables with two rules firing for $\mathcal{C}^+$ and two firing for $\mathcal{C}^-$, $x_i = x_1$ and $x_j = x_2$.

# 3.4 Geometric Characteristics of the Minimum-t-norm

As we have seen, a fuzzy max-min classifier decides locally on the basis of two attributes. It is useful to know how this looks like when considering the geometry of such a classification.

Already when considering only two rules the separation between them may be bended several times depending on the variables that are relevant. But the separations are always parallel to $n-2$ coordinates, because only the two coordinates that are (semi-)relevant at the separation define its direction.

A different situation only occurs at those points, where a (semi-)relevant variable becomes irrelevant and an irrelevant becomes (semi-)relevant. In these points there are more than two relevant variables.

Geometrically these points occur as a bend in the separations.

**Example 7**

*We consider the data space $\mathbb{R}^3$. The class $\mathcal{C}^+$ is defined by $x + y \geq 1$ and $y + z \geq 1$ as to be seen in figure 3.11. At the points of the line passing through $(0, 1, 0)$ and $(1, 0, 1)$, all three variables are relevant.*

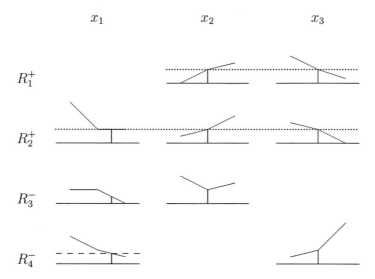

Figure 3.9: The example after having pushed $x_1$. $R_1^+$ and $R_2^+$ did not change, the firing degree of $R_4^-$ decreased with $x_3$ not being dominating any more, and because of $\mu_{R_3^-} < \mu_{R_4^-}$ $R_3^-$ is not firing any more.

## 3.5   Consequences

In this chapter we innovated a generalization of the result of [46], that often only approximate solutions can be constructed, when crisp sets are used. We have shown that fuzzy max-min classification systems assign data to a class locally on the basis of mainly two attributes. The set of points for which this property is satisfied, is a dense set within the class boundaries. Although this sounds like a negative result, it has also positive aspects. First of all, the result holds only locally so that the classification system can still take all attributes into account, when we consider it from a global point of view. And although the local reduction to two variables seems to be very restrictive, it is positive in terms of interpretability. Since we usually want interpretable fuzzy rules, this property definitely helps to understand the rules – especially when we take into account that humans usually do not consider a larger number of attributes simultaneously. Our result can also be applied to analyze a fuzzy max-min classification system, i.e. which attributes are relevant in which region.

It should also be noted that we can at least approximate any kind of (con-

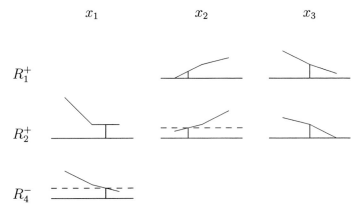

Figure 3.10: The example after having pushed $x_2$. The firing degree of $R_1^+$ and $R_2^+$ is given by $x_2$. Because of $\mu_{R_1^+} < \mu_{R_2^+}$ $R_1^+$ is nor firing any more. We have $\mu_{c^+} = \mu_{R_2^+} = \mu_{R_4^-} = \mu_{c^-}$.

tinuous) class boundaries by fuzzy max-min classification systems and that if we replace the maximum or minimum by another t-conform or t-norm, the situation is completely different [46]. E.g. it turns out that any piecewise linearly separable problem can be solved if we use the Łukasiewicz-t-norm. The method of construction will be demonstrated in the following chapter and generalized in chapter 7.

In chapter 4, we will demonstrate how to construct a fuzzy classification system in the two-dimensional case that uses the Łukasiewicz-t-norm. We have seen in this chapter, that a fuzzy classification system with the Minimum-

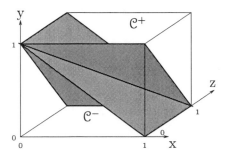

Figure 3.11: An example for a separation with more than two relevant variables.

t-norm would be able to solve the problem in the 2-dimensional case, but already with 3-dimensional data, it would not be possible to solve an arbitrary piecewise linearly separable problem.

Therefore, we concentrate on the Lukasiewicz-t-norm and take a closer look at it by examining its geometric characteristics. Chapter 7 provides, construction for the n-dimensional case, so that finally we will be able to build a fuzzy classification system that solves any piecewise linearly separable system.

# Chapter 4

# Fuzzy Classification Systems and the Łukasiewicz-t-Norm

As fuzzy classification systems usually aim at describing a function from continuous domains – the given attributes – to a discrete domain – representing the classes –, it usually does not make sense to interpolate between the discrete classes. Therefore, one of the main issues for fuzzy classification systems is the question whether different classes can be distinguished by such a fuzzy system.

In this chapter we discuss the question how complicated fuzzy classification rules have to be in order to distinguish classes that are separated by a number of (hyper-)planes. Our intention is to build a fuzzy classifier with rules that are as comprehensible as possible.

We do not aim at a learning algorithm for fuzzy classifiers, but we want to build a fuzzy classification system that provides the same classification as an existing classifier, a neural network for instance.

We restrict our investigations to the case of two classes. Nevertheless, our results can also be applied for a larger number of classes, since we can simply consider the separation of one class w.r.t. the union of all other classes. First we concentrate on the two-dimensional case. It turns out that even in the three-dimensional case the rules can become quite complicated.

The two-dimensional case reveals the principles of the fuzzy classification system with the Łukasiewicz-t-norm. Therefore, investigating the two - dimensional case in this chapter first, it will be easier to understand the construction for the n-dimensional data space as it will be discussed in chapter 7.

In this chapter we will demonstrate how to solve a piecewise linearly separable problem using the Łukasiewicz-t-norm. We will start with the linearly separable case and end with a construction method for arbitrary linearly separable classifications.

## 4.1  Fuzzy Classification with Łukasiewicz-t-norm

Our intention is to build a fuzzy classification system with convenient fuzzy sets. In this chapter we will approach step by step such a system. The fuzzy sets should be triangular and the rules easy to survey.

We consider a set of real attributes similar to the input variables in fuzzy control and a finite number of classes. Each class represents a subspace in the product space of the attribute variables.

The fuzzy classification rules are of the form

If $x_1$ is $\mu_1$ and ... and $x_n$ is $\mu_n$ then the class is $C$.

A classical approach is to use the minimum to evaluate the AND, but it was shown in chapter 3 that, in this case when the AND is evaluated by the minimum, the rules pointing to a class are aggregated by the maximum and an element is assigned to the class with highest membership degree, fuzzy classification systems decide locally only on the basis of two variables. Thus these systems can solve a three-dimensional linearly separable classification problem only approximately.

When the minimum used is replaced by the Łukasiewicz t-norm arbitrary linearly separable problems can be solved by a fuzzy classification system [46]. The shape of class boundaries depending on different t-norms was discussed by Nürnberger et al. [79].

The fuzzy set of rule $R$ for the $i^{th}$ variable is denoted by $\mu_R^{(i)}$, and the firing degree of the rule $R$ is denoted by $\mu_R$. We calculate the firing degree by the Łukasiewicz-t-norm:

$$\mu_R(a_1, ..., a_n) = \max\{0, \sum_{i=1}^{n} \mu_R^{(i)}(a_i) + 1 - n\}.$$

Here we see, that a rule produces a membership degree greater than zero either above or below a hyperplane that is defined by

$$\sum_{i=1}^{n} \mu_R^{(i)}(a_i) = n - 1,$$

because the $\mu_R^{(i)}$ are linear. Such an interpretation of the system as a classification classifying by hyperplanes always holds, when the membership functions are piecewise linear.

The membership degree of the rule increases perpendicularly to this hyperplane.

In the two-dimensional case as shown in figure 4.2, we can construct a classifier, that fulfills the following: For each point $p$ of the data space and for each class there is only one rule firing in $p$. Therefore, we do not have to consider a t-conorm in this case.

According to chapter 3, $\mathcal{C}^+$ and $\mathcal{C}^-$ denote the two classes and $\mathcal{D}$ the area that does not contain any classified point, neither for $\mathcal{C}^+$ nor for $\mathcal{C}^-$. $\mathcal{D}$ is called separation line resp. separation area.

The following notations summarize these definitions:

$$
\begin{aligned}
(x,y,z) \in \mathcal{D} \quad &\Leftrightarrow \quad \mu_{\mathcal{C}^+}^{(\mathcal{R})}(x,y,z) = \mu_{\mathcal{C}^-}^{(\mathcal{R})}(x,y,z), \\
(x,y,z) \in \mathcal{C}^+ \quad &\Leftrightarrow \quad \mu_{\mathcal{C}^+}^{(\mathcal{R})}(x,y,z) > \mu_{\mathcal{C}^-}^{(\mathcal{R})}(x,y,z), \\
(x,y,z) \in \mathcal{C}^- \quad &\Leftrightarrow \quad \mu_{\mathcal{C}^+}^{(\mathcal{R})}(x,y,z) < \mu_{\mathcal{C}^-}^{(\mathcal{R})}(x,y,z).
\end{aligned}
$$

## 4.2 Piecewise Linear Separation Lines

This section will deal with the case of a piecewise linear separation line, that can be described by a function. First we just consider the domain $[a_1, b_1] \times [a_2, b_2]$ with one linear separation as to be seen in figure 4.1 with $\mathcal{C}^-$ above and $\mathcal{C}^+$ below the separation line. The other case can be treated analogously.

We define two rules $R^+$ and $R^-$ with $\mu_{R^+}^{(1)}, \mu_{R^-}^{(1)}, \mu_{R^+}^{(2)}$ and $\mu_{R^-}^{(2)}$ as following:

$$
\begin{aligned}
R^+ &: \text{ If } x_1 \text{ is } \mu_{R^+}^{(1)} \text{ and } x_2 \text{ is } \mu_{R^+}^{(2)} \text{ then class is } \mathcal{C}^+ \\
R^- &: \text{ If } x_1 \text{ is } \mu_{R^-}^{(1)} \text{ and } x_2 \text{ is } \mu_{R^-}^{(2)} \text{ then class is } \mathcal{C}^-
\end{aligned}
$$

with

$$
\begin{aligned}
\mu_{R^+}^{(1)}(x) = \tfrac{x-a_1}{b_1-a_1} \quad &\text{and} \quad \mu_{R^-}^{(1)}(x) = \tfrac{b_1-x}{b_1-a_1} = 1 - \mu_{R^+}^{(1)}(x), \\
\mu_{R^+}^{(2)}(y) = \tfrac{b_2-y}{b_2-a_2} \quad &\text{and} \quad \mu_{R^-}^{(2)}(y) = \tfrac{y-a_2}{b_2-a_2} = 1 - \mu_{R^+}^{(2)}(y).
\end{aligned}
$$

Then the two classes are correctly defined. Figures 4.1 and 4.2 are to illustrate how the rules work. Figure 4.1 shows the separating line and figure 4.2 demonstrates the firing degree of the two rules.

Let the data space be $[a_1, b_1] \times [a_2, b_2]$. In the case of having exactly one point $(x,y)$ for each $x$ that belongs to the separation line, we can describe the separation line by a function $f : [a_1, b_1] \to [a_2, b_2]$, $f(x) = y$ with $(x,y) \in \mathcal{D}$. With this we define

$$
a := \min \left\{ y \in [a_2, b_2] \mid \exists x \in [a_1, b_1] : (x,y) \in \mathcal{D} \right\} = \min_{x \in [a_1, b_1]} f(x),
$$

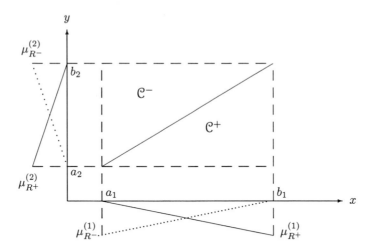

Figure 4.1: The fuzzy sets used for the classification

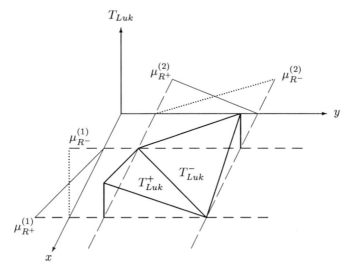

Figure 4.2: An illustration of the firing degrees of the rules using $T_{Luk}$

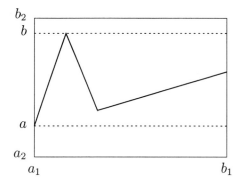

Figure 4.3: An example for determining the values of range for a separation function

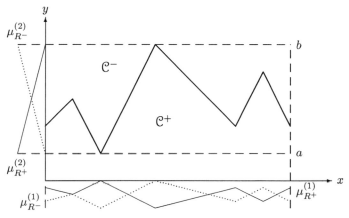

Figure 4.4: Illustration of the fuzzy sets for a piecewise linear separation

$$b := \max \{ y \in [a_2, b_2]; \mid \exists x \in [a_1, b_1] : (x, y) \in \mathcal{D} \} = \max_{x \in [a_1, b_1]} f(x).$$

Note that $a_2 \leq a \leq b_2$ and $a_2 \leq b \leq b_2$, but not necessarily $a, b \in \{a_2, b_2\}$. Figure 4.3 shows an example for such a case.

Then we define

$$\mu_{R-}^{(2)}(y) := \tfrac{y-a}{b-a} \quad \text{and} \quad \mu_{R+}^{(2)}(y) := \tfrac{b-y}{b-a} \ . \tag{4.1}$$

$$\mu_{R-}^{(1)}(x) := \tfrac{b-f(x)}{b-a} \quad \text{and} \quad \mu_{R+}^{(1)}(x) := \tfrac{f(x)-a}{b-a} \ . \tag{4.2}$$

We have a proportional relation between $\mu_{R+}^{(1)}$ and $f$, if we consider $a$ to be the "zero-line", and a reciprocally proportional one between $\mu_{R-}^{(1)}$ and $f$.

**Remark 6**

*Equations (4.1) and (4.2) confirm that this method also works for each kind of continuous separation, that can be described by a function $f : \mathbb{R} \to \mathbb{R}$. We simply have to construct $\mu_{R^-}^{(1)}$ and $\mu_{R^+}^{(1)}$ by using $f$.*

**Remark 7**

1. *When calculating $\mu_{R^+}$ and $\mu_{R^-}$, we can see that this method considers, whether the point is above the separation line $(y > f(x))$ or below $(y < f(x))$, when $f$ describes the separation line itself.*

2. *In every rule there is either $\mu_{R^+}(x,y) = 0$ (if $f(x) \geq y$) or $\mu_{R^-}(x,y) = 0$ (if $f(x) \leq y$). Therefore we just have a membership degree not equal to 0 for the class where the considered point is located in, and this membership is generated by one single rule. Iff $(x,y) \in \mathcal{D}$, then the membership degree for $(x,y)$ for $\mathcal{C}^+$ and for $\mathcal{C}^-$ is $\mu_{R^+}(x,y) = \mu_{R^-}(x,y) = 0$.*

Figure 4.5 shows an example for a separation that can not be described by a single function. Therefore we have to divide the problem into smaller ones that can be solved by constructing such a function.

We divide the space into disjoint rectangles with sides parallel to the axis, so that each rectangle $R = A \times B$, $A, B \subset \mathbb{R}$ being intervals, has a separation that can be described by a function. To reduce the problem to easily manageable ones, we even assume that the separations are lines without bends inside one of these rectangles.

The easiest way to evaluate the rectangles is the following: We calculate all the intersection points of the separation lines inside the data space. Then we draw horizontal and vertical lines through these points. Finally we have a grid with rectangles that are of three types of rectangles that will be discussed in the following.

In section 7.2, we will explain another method, that leads to a lower number of rectangles and consequentially to a lower number of rules, but here we concentrate on constructing the rules.

**Definition 8 (a)**

*A rectangle $A \times B$ is called rectangle of type 0 if it does not contain any point of $\mathcal{D}$ as an inner point, so the whole rectangle belongs to one class exept for the points of the boundary of the rectangle, that can also belong to $\mathcal{D}$.*

*A rectangle $A \times B$ is called rectangle of type 1, if for every value $x \in A$ there is exactly one $y \in B$ so that $(x,y) \in \mathcal{D}$.*

As a rectangle of type 0 completely belongs to one class, we only need one rule to classify the points inside. A rectangle of type 1 is treated as described

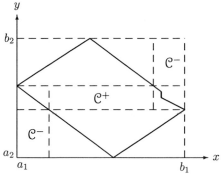

Figure 4.5: An example for the division of the space into rectangles

so far, so that we obtain two rules firing inside the rectangle.

If we have a line, that is parallel to the second coordinate, we can still use the method for a rectangle of type one, but now we use a function $f' : \mathbb{R} \to \mathbb{R}$ and change the definitions for $x$ and $y$:

$$\mu_{R^-}^{(1)}(x) := \frac{x-c}{d-c} \text{ and } \mu_{R^+}^{(1)}(x) := \frac{d-x}{d-c}, \tag{4.3}$$

$$\mu_{R^-}^{(2)}(y) := \frac{d-f(y')}{d-c} \text{ and } \mu_{R^+}^{(2)}(y) := \frac{f'(y)-c}{d-c} \tag{4.4}$$

$$c := \min \{ x \in [a_1, b_1] \mid \exists y \in [a_2, b_2] : (x,y) \in \mathcal{D}\} = \min_{y \in [a_2,b_2]} f'(y),$$

$$d := \max \{ y \in [a_2, b_2]; \mid \exists x \in [a_1, b_1] : (x,y) \in \mathcal{D}\} = \max_{y \in [a_2,b_2]} f'(y).$$

In the example of figure 4.5 we need seven rectangles. The small one on the right side is an example for a rectangle containing a vertical line.

**Remark 8**

*We aim at having interpretable fuzzy sets. Therefore we should choose the rectangles either that way that the separation line is the diagonal of the rectangle or that the separation line is triangular in the rectangle. Then the fuzzy sets are also linearly decreasing or increasing or they are triangular.*

For this reason, we should divide the rectangle in figure 4.5 with the vertical line into six rectangles:

In appendix B.1 the construction for the calculation can be found in a summarized form. You will find the algorithm for the rectangle of type 0 on page 159 and for the rectangle of type 1 on page 160

## 4.3    Treatment of the Acute Angle

Assume that the slopes of the functions $f$ and $g$ as described in the previous section define the two sides of an angle and have the same sign. Then we have a rectangle with two separation lines instead of one. Figure 4.6 shows an example. The separation lines $s_1$ and $s_2$ are described by the functions $f$ and $g$.

**Definition 8 (b)**
*A rectangle with two separation lines that meet in one corner of the rectangle and form an acute angle a rectangle of type 2.*

When we talk of an "acute angle", we mean the angle of the two separating lines inside a rectangle of type 2. Here the previously explained procedure does not work, so this case has to be treated differently.

Such a rectangle that can not be partitioned into rectangles of type 0 and type 1 to determine each separation separately. We have to find another solution. In the following we will suggest two possiblities. In the first suggestions, we will choose a negligible small rectangle containing the acute angle and accept a certain misclassification.

The other way of construction is accurate. We choose the two rules for one separation as described in the previous sections. Additionally we will introduce a third rule that starts between the two separation lines and overtakes the rule classfying the inner part of the acute rectangle exactly at the second separation line, so that we have a correct classification outside the acute angle.

**Remark 9**
*It is always possible to choose the rectangle that way, that the line with the larger absolute slope is the diagonal of the rectangle, while the other line reaches the side of the rectangle at $c_2 = a_2 + \beta \cdot (b_2 - a_2)$. Then $\beta$ is the factor between the two slopes.*

We have two possibilities to treat the acute angle. The first one is more practical, admitting some wrong-classified points, while the second one results in a proper distinction between the classes but produces rules that are difficult to interpret.

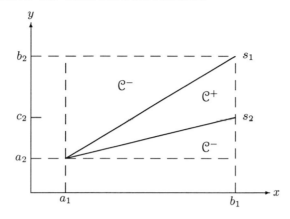

Figure 4.6: An acute angle between the separation lines $s_1$ and $s_2$

## 4.3.1 First Possibility: Negligible Small Rectangle

If the rectangle of type 2 is small enough, then it is possible to accept the misclassification that occurs if we simply ignore one or both separations inside the rectangle.

But the smaller we choose the rectangle containing the acute angle, the smaller the neighboring rectangles have to be chosen. Then we need more rectangles to cover the whole space. The fitting of the rectangles can be seen in figure 4.7.

We have to choose the size of the rectangle containing the acute angle in a way that we can except the misclassification $\alpha$ but still do not have too many rectangles.

**Algorithm 1 (Rectangle of Type 2, neglecting small rectangle)**

*choose small neglectable rectangle;*
*Construct one rule for this rectangle;*
*while end of data space not reached*
    *do construct next rectangle of type 1 in direction of x;*
        *construct rules for this rectangle;*
        *construct rectangle of type 1 above for second line;*
        *construct rules for this rectangle;*
*add the rules to the rulebase;*

◇

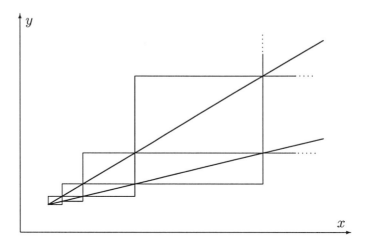

Figure 4.7: The construction of the rectangle , starting with the rectangle of type 2 and continuing with rectangles of type 1

The exact calculation for the algorithm can be found in appendix B.1 on page 161.

## 4.3.2 Second Possibility: Using uncommon Rules

It is possible to construct rules that solve the problem exactly, but than we have to accept that some of the fuzzy sets do not reach membership degree 1.

We are going to construct three rules to classify the points of the rectangle correctly: Two rules start at one separation increasing into opposite directions. This classifies correctly one outer section and the inner section of the acute angle. To classify the second outer part correctly, we add another rule, that starts later, but increases faster than the rule classifying the inner part of the acute angle. It overtakes the other rule exactly ar the second separation line. By this method we get a correct classification. Now we will contruct the formula for this method:

For the visualization of the rules in this section we use a different way of plotting the firing degrees of the rules: The horizontal axis is the line from $(a_1, b_2)$ to $(b_1, a_2)$, and the vertical axis shows the firing degree of the rules. In figure 4.8 we see, how this looks, when we take the two rules for each separation line as we did in section 4.2 without considering that they influence each other. The area above $s_1$ and the area below $s_2$ is misclassified for $\mathcal{C}^+$ instead of $\mathcal{C}^-$.

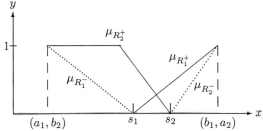

Figure 4.8: Two rules for each separation line in the rectangle

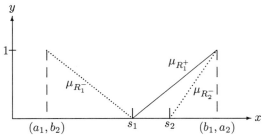

Figure 4.9: After removing the rule $\mu_{R_2^+}$.

We can start to solve this problem by removing one rule for $\mathcal{C}^+$. The result can be seen in figure 4.9, where the area below $s_2$ is still misclassified for $\mathcal{C}^+$. If we removed $\mu_{R_1^+}$, too, then we would not have a rule at all firing in the area between the two separation lines, so that this section would belong to $\mathcal{D}$ instead of $\mathcal{C}^+$. Therefore, we keep $\mu_{R_1^+}$.

When the rule $\mu_{R_2^-}$ at the right side cuts the rule $\mu_{R_1^+}$ at the second separation line as to be seen in figure 4.10, then we have a change between the two classes at the separation line $s_2$.

The only problem that is left in this case is the fact that not each of the resulting fuzzy sets $\mu_{R_i}^{(j)}$ satisfies the condition, that there has to be an input-

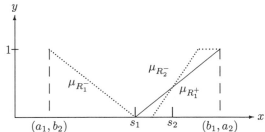

Figure 4.10: An example for a right classification

value $x_i$ so that $\mu_{R_i}^{(j)}(x_i) = 1$. This means that the fuzzy set does not reach membership degree 1 in every case.

Now we turn towards the concrete construction of these rules: For the two rules $\mu_{R_1^+}$ and $\mu_{R_1^-}$, that are taken from the basic constructions as shown in figure 4.8, the definition is well known:

$$\mu_{R_1^-}^{(1)}(x) = 1 - \frac{x-a_1}{b_1-a_1} \quad \text{and} \quad \mu_{R_1^-}^{(2)}(y) = \frac{y-a_2}{b_2-a_2},$$
$$\mu_{R_1^+}^{(1)}(x) = \frac{x-a_1}{b_1-a_1} \quad \text{and} \quad \mu_{R_1^+}^{(2)}(y) = 1 - \frac{y-a_2}{b_2-a_2},$$

and therefore

$$\mu_{R_1^-}(x,y) = \max\{0, \frac{2-a_2}{b_2-a_2} - \frac{x-a_1}{b_1-a_1}\} \quad \text{and}$$
$$\mu_{R_1^+}(x,y) = \max\{0, \frac{x-a_1}{b_1-a_1} - \frac{y-a_2}{b_2-a_2}\}.$$

After having removed $\mu_{R_2^+}$ and the old $\mu_{R_2^-}$, we have the new rule $\mu_{R_2^-}$ that fits in with $\mu_{R_1^-}$ and $\mu_{R_1^+}$, if

$$\mu_{R_2^-}(x,y) = \mu_{R_1^+}(x,y) \quad \Leftrightarrow \quad (x,y) \in s_2 \qquad \text{and} \qquad (4.5)$$

$$\mu_{R_2^-}(x,y) = 0 \qquad \Leftrightarrow \quad y = \frac{b_2 + c_2 - 2 \cdot a_2}{2 \cdot (b_1 - a_1)} \cdot (x - a_1) + a_2. \quad (4.6)$$

The first condition guarantees that $\mu_{R_2^-}$ overtakes $\mu_{R_1^+}$ at $s_1$, while the second condition lets $\mu_{R_2^-}$ start with firing degree 0 at the bisecting line of $s_1$ and $s_2$. This is only one possibility to construct $\mu_{R_2^-}$. It is possible to start $\mu_{R_2^-}$ at any line between $s_1$ and $s_2$ that meets the two lines in their intersection point.

When $\mu_{R_2^-}$ is described with condition (4.6), then it increases twice as fast as $\mu_{R_1^+}$.

**Corollary 3**

*The necessary conditions (4.5) and (4.6) for the construction of $\mu_{R_2^-}$ are fulfilled with*

$$\mu_{R_2^-}(x,y) = \max\{0, (1+\beta) \cdot \frac{x-a_1}{b_1-a_1} - 2 \cdot \frac{y-a_2}{b_2-a_2}\}$$

*with $\beta = \frac{c_2-a_2}{b_2-a_2} \in ]0; 1[$. The fuzzy sets for this rule are*

$$\mu_{R_2^-}^{(1)}(x) = (1+\beta) \cdot \frac{x-a_1}{b_1-a_1} \quad \text{and} \quad \mu_{R_2^-}^{(2)}(y) = 1 - 2 \cdot \frac{y-a_2}{b_2-a_2}.$$

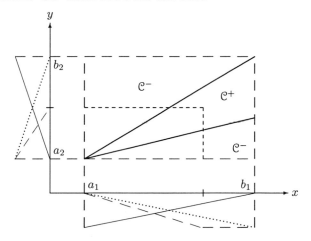

Figure 4.11: Here the classification only works inside the inner rectangle.

**Proof:** Inside the rectangle, $\mu_{R_1^+}$ is described by $\mu_{R_1^+}(x,y) = \frac{x-a_1}{b_1-a_1} - \frac{y-a_2}{b_2-a_2}$ and $s_2$ is described by

$$y = \frac{x-a_1}{b_1-a_1} \cdot (c_2 - a_2) + a_2.$$

Therefore we can proof (4.5) by

$$
\begin{aligned}
\mu_{R_1^+}(x,y) &= \mu_{R_2^-}(x,y) \\
\Leftrightarrow \quad \frac{x-a_1}{b_1-a_1} - \frac{y-a_2}{b_2-a_2} &= (1+\beta) \cdot \frac{x-a_1}{b_1-a_1} - 2 \cdot \frac{x-a_1}{b_2-a_2} \\
\Leftrightarrow \quad \frac{y-a_2}{b_2-a_2} &= \beta \cdot \frac{x-a_1}{b_1-a_1} \\
\Leftrightarrow \quad y &= \frac{c_2-a_2}{b_1-a_1} \cdot (x-a_1) + a_2
\end{aligned}
$$

The second condition (4.6) can easily be shown by

$$
\begin{aligned}
\mu_{R_2^-}(x,y) &= (1+\beta) \cdot \frac{x-a_1}{b_1-a_1} - 2 \cdot \frac{y-a_2}{b_2-a_2} = 0 \\
\Leftrightarrow \quad \frac{2(y-a_2)}{b_2-a_2} &= \frac{b_2+c_2-2a_2}{b_2-a_2} \frac{x-a_1}{b_1-a_1} \\
\Leftrightarrow \quad 2(y-a_2) &= (b_2+c_2-2a_2)\frac{x-a_1}{b_1-a_1} \\
\Leftrightarrow \quad y &= \frac{b_2+c_2-2\cdot a_2}{2\cdot(b_1-a_1)} \cdot (x-a_1) + a_2.
\end{aligned}
$$

$\square$

We still have the problem, that the membership degree should stay between 0 and 1. With the above definitions we have $\mu_{R_2^-}^{(2)} > 1$ for $x > \frac{1}{1+\beta} \cdot (b_1 - a_1)$ and $\mu_{R_2^-}^{(2)} < 0$ for $y > \frac{1}{2} \cdot (b_2 - a_2)$. If we simply cut the fuzzy sets at the membership degree 0 resp. 1, then the fuzzy sets would have a bend as

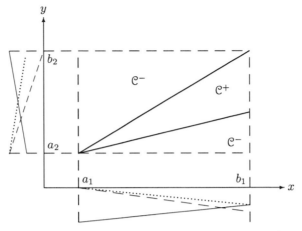

Figure 4.12: After having scaled the rules with the factor $\frac{1}{2}$, ($\mu_{R_1^+}$ continuous line $/\cdots \mu_{R_1^-}/ -- \mu_{R_2^-}$)

demonstrated in figure 4.11 with $\mu_{R_1^+}$ drawn with a continuous line, $\mu_{R_1^-}$ with a dotted and $\mu_{R_2^-}$ with a dashed line. The classification would only be correct inside the small rectangle marked by the dashed lines. Because of $\frac{1}{1+\beta} > \frac{1}{2}$ we scale all the fuzzy sets with $\frac{1}{2}$. This procedure can be considered like choosing the largest possible rectangle that is correctly classified and similar to $[a_1, b_1] \times [a_2, b_2]$ and scaling up this rectangle to the size of the big one. Finally we get

$$\mu_{R_1^+}^{(1)}(x) = 1 - \frac{x-a_1}{2(b_1-a_1)} \text{ and } \mu_{R_1^+}^{(2)}(y) = \frac{y-a_2}{2(b_2-a_2)},$$
$$\mu_{R_1^-}^{(1)}(x) = \frac{x-a_1}{2(b_1-a_1)} \text{ and } \mu_{R_1^-}^{(2)}(y) = 1 - \frac{y-a_2}{2(b_2-a_2)}$$

for the basic rules that belong to the classification line $s_1$ and

$$\mu_{R_2^-}^{(1)}(x) = \frac{1+\beta}{2} \cdot \frac{x-a_1}{b_1-a_1} \text{ and } \mu_{R_2^-}^{(2)}(y) = 1 - \frac{y-a_2}{b_2-a_2}$$

for the additional rule that is responsible for the second classification line.

These final rules as shown in figure 4.12 draw a correct classification, but $\mu_{R_1^-}(x_1)$ and $\mu_{R_2^-}(x_1)$ do not reach membership degree 1.

The construction for the straight forward calculation of these rules can be found in appendix B.1 on 162.

### 4.3.3 Generalization for the Acute Angle

In 4.3.2 we gave an exact construction for a possible representation of the classifier, but there exist other solutions. The constructed rules only need to fulfill the following conditions:

- The two rules $R_1^-$ and $R_2^-$ are constructed as in 4.2, but they can start with firing degree 0 at the classification lines themselves or at lines between the two. We do not need to use the same line for both rules.

- $R^+$ is chosen that way that it has the same membership degree as $R_1^-$ resp. $R_2^-$ at the classification lines $s_1$ and $s_2$.

- During the construction the fuzzy sets are not cut at 1. Scaling leads towards the final rules that yield membership degrees with the values between 0 and 1.

**Remark 10**
*We can realize by the following argument that it is impossible in general that all the fuzzy sets reach 1:*
*The rule for $\mathcal{C}^+$ has to cut at least one rule for $\mathcal{C}^-$ at $s_1$ or $s_2$. So it has to be more steep than the other rule. We can just scale the two together, and if we increase both so that the fuzzy sets for the lower rule reach one, the fuzzy sets for the other rules are cut and therefore do not work anymore in the area, where they are cut.*

**Remark 11**
*For the construction we considered the case with the class $\mathcal{C}^+$ between the two separation lines and $\mathcal{C}^-$ outside this sector. The construction also works in the multiclass case when e.g. we have the class $\mathcal{C}_1$ above $s_1$, $\mathcal{C}_2$ between $s_1$ and $s_2$ and $\mathcal{C}_3$ below $s_2$. Then $\mu_{R_{e_1}} = \mu_{R_1^-}$, $\mu_{R_{e_2}} = \mu_{R_1^+}$ and $\mu_{R_{e_3}} = \mu_{R_2^-}$.*

## 4.4 Improper Separation Lines

In practice, we often do not have proper separation lines, but separation areas not containing any classified points. An example for this case can be seen in figure 4.13 where the unclassified area is hatched. Then the situation becomes much easier as we have more scope to decide how to divide the space into several rectangles.
Figure 4.13 demonstrates that choosing the rectangles in a proper way can lead to a quite simple solution with three rectangles of type 0 and type 1 avoiding the most complicated case of a rectangle of type 2.

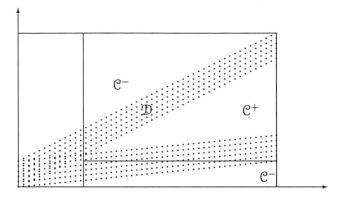

Figure 4.13: Choosing the rectangles cleverly

If we would simply follow the unclassified regions and consider the whole problem in one rectangle, it would be the complicated case of the acute angle, but we are free to consider the simple case of three rectangles, i.e. two of type 0 and one of type 1.

## 4.5   Consequences

In this chapter we have seen a method to construct a Łukasiewicz-t-norm-based fuzzy classification system that describes a piecewise linear classification system in the two-dimensional case.

In the two-dimensional case, we have constructed the rules in such a way that they do not overlap for single classes, so that there is no need of rule aggregation, and considering t-conorms becomes superfluous.

In the next chapter we will consider the other direction. This means that we examine how to derive the geometric characterization of a classification from a given fuzzy classification system.

After the following chapter we will have a good understanding of the different principles that we use thoughout this work. Then we can turn towards higher dimensions and apply the principles to multi-dimensional classification problems.

We are not restricted to the two-dimensional case. As there is a general way to construct a fuzzy classification system with the Łukasiewicz-t-norm for any piecewise linear classification problem in any dimension, we will give the algorithm for this in chapter 7.

# Chapter 5

# Geometric Characterization of a Fuzzy Classification System

In chapter 3, we have seen the restrictions of a max-min-classifier. As consequence, we have used the Łukasiewicz-t-norm. When we are given a fuzzy classification system, it would be interesting to know how the classification looks like from the geometrical point of view. To determine these geometric characterizations, we have to visualize the rules [93]. In this chapter we are examining the characterization of a fuzzy classification system.

We will start by determining how far the fuzzy sets of the rules should overlap in each dimension to receive a convenient geometric characterization and then examine how several rules interfere with each other. For this we first restrict our investigations to one subcuboid of the data space and then extend the results for the whole data space.

## 5.1 Basic Requirements

In this section we want to examine how to devide the data space into cuboids that can easily be handled. First of all we show how the separations of the data space depend on fuzzy sets when the fuzzy sets overlap for each single dimension. By this we get an overview on how the data space can be devided into reasonable multi-dimensional cuboids.

We assume that for each dimension $i$ there are $z_i$ fuzzy sets $\mu_1^{(i)}, \ldots, \mu_{z_i}^{(i)}$ with the center $c_j^{(i)}$ as shown in figure 5.1.

Furthermore we assume that the fuzzy sets are uniformly distributed in each dimension. This means that

$$\mu_j^{(i)}(x_i) = \max\{0, 1 - \gamma_i \cdot |x_i - c_j^{(i)}|\} \tag{5.1}$$

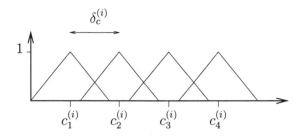

Figure 5.1: Regularly distributed fuzzy sets for dimensions i

holds for $\gamma_i$ being the same for all fuzzy sets for the $i^{th}$ coordinate and $c_j^{(i)}$ being the middle of each fuzzy set. There is a value $\delta_c^{(i)}$ so that for two such points $c_j^{(i)}$ and $c_{j+1}^{(i)}$ belonging to neighboring fuzzy sets the equation $c_{j+1}^{(i)} - c_j^{(i)} = \delta_c^{(i)}$ holds. The distribution of the fuzzy sets can be seen in figure 5.1. The degree of overlap that is ideal for calculating the classification will be determined in section 5.2.

Each rule $R$ has for each coordinate $i$ a unique fuzzy set $\mu_j^{(i)}$. Therefore, for the $i^{th}$ dimension there is a function

$$f_i: \quad \mathcal{R} \quad \rightarrow \quad \{1, \ldots, z_i\}$$
$$R \quad \mapsto \quad j.$$

Then the rule $\mu_R$ is defined by $\mu_{R_k}(x_1, \ldots, x_n) = \max\{0, \sum_{i=1}^{n} \mu_{R_k}^{(i)}(x_i) + 1 - n\}$ with $\mu_R^{(i)} = \mu_{f_i(R)}^{(i)}$.

We can divide the space into different cuboids by cutting each coordinate at those points into smaller intervals, where the fuzzy sets reach membership degree 1, this is at the input values $c_j^{(i)}$. In this way we obtain a lattice with regular cuboids as shown in figure 5.2. Note that we can scale the coordinates, so that it is possible to transform the lattice into a lattice with squares instead of arbitrary cuboids.

To characterize the classification in the data space, we will start by considering a single cuboid in section 5.3, and then combine the classification of several cuboids to cover the whole data space as described in section 5.4.

## 5.2   Overlap of Fuzzy Sets

It is important to determine how far the fuzzy sets should overlap. For this reason this section will present a characterization of the rules. Then it will become obvious why $\delta_c^{(i)}$ must be neither too big nor to small.

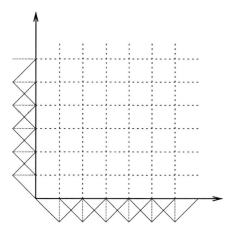

Figure 5.2: The lattice with regular cuboids

**Remark 12**

*When the fuzzy sets overlap in a relatively large region, then also the regions, where the rules fire, overlap. The more the rules overlap, the more rules fire in each point, and this requires expensive calculations.*

*As we aim at a distinction between different classes, a clear decision is desirable. Therefore it is reasonable to have only few rules firing in one point and determining a clear assignment to one of the classes.*

If the value of $\delta_c^{(i)}$ is too big then there is a section in the middle of the cuboid, where no rule fires. Then the points of this section can not be classified. If $\delta_c^{(i)}$ is too small then the sets $S_{R_i}$ where the rules $R_i$ fire overlap too much. This results in a great number of rules firing in the same region. We prefer to determine only a small number of rules for each point.

Let $[a_1, b_1] \times \cdots \times [a_n, b_n]$, $[a_i, b_i] \subset X_i$, be one cuboid of the lattice that covers the data space $X_1 \times \cdots \times X_n \subset \mathbb{R}^n$ (compare figure 5.3). In the following part of this section we assume, that $(b_i - a_i)$ is equal for all $i \in \{1, \ldots, n\}$, so that we have a cube for the cuboid. Any other cuboid with different edges can easily be achieved by scaling.

Figure 5.2 and figure 5.3 demonstrate how the fuzzy sets overlap. We consider the membership degree only inside the chosen cuboid, and there $\mu_{R_j}^{(i)}$ is linear.
As $\mu_{R_j}^{(i)} \geq 0$ holds inside the cuboid, we can write

$$\mu_{R_j}^{(i)}(x_i) = 1 - \gamma_i \cdot (x_i - c_k^{(i)}) \quad \text{resp.} \quad \mu_{R_j}^{(i)}(x_i) = 1 - \gamma_i \cdot (c_k^{(i)} - x_i)$$

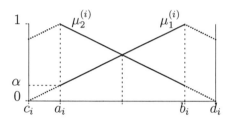

Figure 5.3: Two overlapping fuzzy sets for the $i^{th}$ coordinate

instead of $\mu_{R_j}^{(i)}(x_i) = \max\{0, 1 - \gamma_i \cdot |x_i - c_k^{(i)}|\}$. In figure 5.3, $\mu_1^{(i)}$ is defined
by the second equation, while $\mu_2^{(i)}$ is described by the first one.
First of all we consider a single fuzzy rule $R$. $(x_1, \ldots, x_n) \in \mathbb{R}^n$ is the input
point and $\mu_R^{(i)}$ is the fuzzy set of $R$ for the $i^{th}$ coordinate. Then the firing
degree of $R$ in $(x_1, \ldots, x_n)$ is calculated by

$$\mu_R(x_1, \ldots, x_n) = \max\{0, \sum_{i=1}^{n} \mu_R^{(i)}(x_i) + 1 - n\}.$$

We use this simplified notation until the end of this section. We start by
choosing $\delta_c^{(i)} = 1/\alpha_i = \gamma_i$ for the fuzzy sets as mentioned in equation 5.1 .
In figure 5.4 an example for the three-dimensional space in shown. The
section that includes the corner $P_1$ resp. the corner $P_2$ belongs to class $\mathcal{C}_1$
resp. $\mathcal{C}_2$, while the middle section is not classified at all. The rule starts at
the grey hyperplane with firing degree 0 and increases orthogonally following
the dashed arrow until it reaches firing degree 1 at the corner $P_i$. This means
that all the points of a hyperplane parallel to the grey one have the same
firing degree. The hyperplane representing the firing degree 0 touches the
corners that are exactly one edge away from the corner $P_i$. At these corners
the membership degree of $n - 1$ fuzzy sets is 1, while the $n^{th}$ membership
degree is 0, so that we obtain $\mu_{R_i}(x_1, \ldots, x_n) = ((n - 1) * 1 + 0) + 1 - n = 0$,
if $(x_1, \ldots, x_n)$ belongs to the hyperplane.

## Definition 9
We call a set $S$ of the data space $\mathcal{C}_i$-possible iff

$$\mu_{\mathcal{C}_i}^{\mathcal{R}}(X) > 0 \text{ for all } X \in S.$$

In figure 5.4, we consider an example with two opposite rules: Two rules
$R_1$ and $R_2$ for the classes $\mathcal{C}_1$ and $\mathcal{C}_2$ with $\mu_{R_2}^{(i)}(x_i) = 1 - \mu_{R_1}^{(i)}(x_i)$ for all
$i \in \{1, \ldots, n\}$. They increase in opposite directions. There are two classified
regions with the classes $\mathcal{C}_1$ and $\mathcal{C}_2$ and one unclassified region $\mathcal{U}$ between.

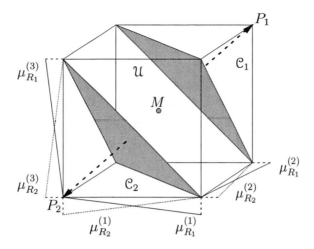

Figure 5.4: An example for the classification with two rules

When the dimension is increasing, then the volume of the unclassified region compared to the volume of the classified ones also increases, so that a big part of the data remains unclassified. Even if we have a complete rule base, there is still an unclassified region around $M = ((a_1 + b_1)/2, \ldots, (a_n + b_n)/2)$. When the fuzzy sets overlap more than in this example, then the separating hyperplanes, that are marked gray in figure 5.4, move towards each other. By this we can diminish the unclassified region.

The separating hyperplanes meet in the middle of the cube, if the fuzzy sets are pushed towards each other by $\alpha \cdot (b_i - a_i)$ with $\alpha = \frac{n-2}{n}$. This can be seen in figure 5.3 with $\alpha = \frac{a_i - c_i}{b_i - c_i}$. The same $\alpha$ is the membership degree of the fuzzy sets at the boundary of the cube.

## Corollary 4
*The fuzzy sets*

$$\mu_1^{(i)}(x_i) = 1 - (b_i - x_i) \cdot \frac{1-\alpha}{b_i - a_i} = 1 - \frac{b_i - x_i}{b_i - c_i} \quad \text{and}$$
$$\mu_2^{(i)}(x_i) = 1 - (x_i - a_i) \cdot \frac{1-\alpha}{b_i - a_i} = 1 - \frac{x_i - a_i}{b_i - c_i}$$

*fulfill the following conditions:*

1. *The separating hyperplanes contain the only points inside the cube that do not have membership degree greater than 0.*

2. *The rules have firing degree 0 in the middle point $M$.*

3. *To each rule we can assign a corner of the cube where the rule has firing degree 1.*

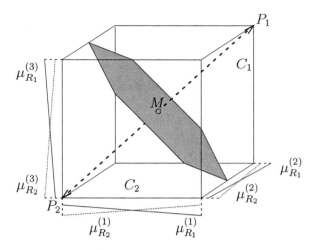

Figure 5.5: How two opposite rules look like when pushed towards each other by $\alpha \cdot (b_i - a_i)$

**Proof:** It is $1 - \alpha = (b_i - c_i - a_i + c_i)/(b_i - c_i) = (b_i - a_i)/(b_i - c_i)$ and therefore $(1 - \alpha)/(b_i - a_i) = 1/(b_i - c_i)$. With this the right part of the equations is easy to calculate.

Note that $1 - \alpha = 1 - \frac{n-2}{n} = \frac{2}{n}$. Then for a rule $R$ we get in the middle-point $M$ of the cube the firing degree

$$
\begin{aligned}
\mu_R(M) &= \sum_{i=1}^{n} \mu_R^{(i)}\left(\frac{a_i - b_i}{2}\right) + 1 - n \\
&= \sum_{i=1}^{n}\left(1 - \frac{b_i - a_i}{2} \cdot \frac{1-\alpha}{b_i - a_i}\right) + 1 - n \\
&= n - \sum_{i=1}^{n} \frac{1-\alpha}{2} + 1 - n \\
&= 1 - \frac{n}{2} - \frac{n}{2} \cdot \frac{n-2}{n} = 0
\end{aligned}
$$

It does not matter whether the $\mu_R^{(i)}$ is increasing until it reaches 1 in $b_i$ or decreasing from 1 in $b_i$, because in the middle $m_i = (a_i + b_i)/2$ the increasing and the decreasing function meet and both sorts of fuzzy sets yield $\mu_R^{(i)}(m_i) = 1 - \frac{b_i - a_i}{b_i - c_i} = \frac{1+\alpha}{2}$. Therefore in $M = (m_1, \ldots, m_n)$ the firing degree is $\mu_R(M) = \sum_{i=1}^{n}\left(\frac{1+\alpha}{2}\right) + 1 - n = n \cdot \frac{\alpha-1}{2} + 1 = n \cdot \left(-\frac{1}{n}\right) + 1 = 0$

The vector from M towards $P_1$ is $((b_1 - a_1)/2, \ldots, (b_n - a_n)/2)$ and turns into the vector $e = (1/\sqrt{n}) \cdot (1, \ldots, 1)$ when being normalized, because $b_i - a_i = b_1 - a_1$ for all $i = 1, \ldots, n$. The hyperplane that includes $M$ and that is orthogonal towards this vector $n_M$ is defined by

$$
e \cdot X - e \cdot M = e \cdot X - \frac{1}{2\sqrt{n}} \sum_{i=1}^{n}(a_i + b_i) = 0. \tag{5.2}
$$

$e$ is the normal vector and $M$ is used to calculate the distance $\frac{1}{2\sqrt{n}}\sum_{i=1}^{n}(a_i + b_i)$ from the hyperplane to 0. For all the points of this hyperplane the firing degree of the rule $R_1$ is

$$
\begin{aligned}
\mu_1(X) &= \sum_{i=1}^{n}\mu_1^{(i)}(x_i) + 1 - n \\
&= \sum_{i=1}^{n}(1 - \frac{2}{n}\cdot\frac{b_i-x_i}{b_i-a_i}) + 1 - n &= 1 - \frac{2}{n}\sum_{i=1}^{n}\frac{b_i-x_i}{b_i-a_i} \\
&= 1 - \frac{2}{n\cdot(b_1-a_1)}(\sum_{i=1}^{n}b_i - \sum_{i=1}^{n}x_i) \\
&\overset{(5.2)}{=} 1 - \frac{2}{n\cdot(b_1-a_1)}(\sum_{i=1}^{n}b_i - \frac{1}{2}\sum_{i=1}^{n}(a_i + b_i)) \\
&= 1 - \frac{2}{n\cdot(b_1-a_1)}(\sum_{i=1}^{n}\frac{b_i-a_i}{2}) &= 1 - (\sum_{i=1}^{n}\frac{2}{n}\cdot\frac{b_i-a_i}{2\cdot(b_i-a_i)}) \\
&= 0
\end{aligned}
$$

Therefore the rule starts with firing degree 0 at a hyperplane including $M$ and being orthogonal towards the vector $(P_1 - M)$, and increases orthogonally towards $P_1$. $\square$

The same holds for each rule $R_i$: the firing degree equals 0 at the hyperplane including $M$ and orthogonal to $(P_i - M)$ with $P_i$ being the point where the fuzzy sets of the rule all have the firing degree 1. Figure 5.5 shows an example with two rules.

Note that a rule in a cuboid that is not a cube still increases from a plane including $M$ toward the corner, but it does not do so orthogonally.

**Remark 13**
Let us assume that we have two fuzzy sets $\mu_{R_1}^{(i)}$ and $\mu_{R_2}^{(i)}$ for the $i^{th}$ coordinate that overlap as shown in figure 5.3.
With $\mu_{R_1}^{(i)}(x_i) = (x_i - c_i)/(b_i - c_i)$ and $\mu_{R_2}^{(i)}(x_i) = 1 - (x_i - a_i)/(b_i - c_i)$, we have the following relation:

$$\mu_{R_2}^{(i)}(x_i) = 1 + \alpha - \mu_{R_2}(x_i))$$

that holds for $x_i \in [a_i, b_i]$.

We return to our example with the two opposite rules as described on page 62. We have $\alpha = \frac{a_i-c_i}{b_i-c_i} = \frac{n-2}{n}$, so that we can calculate the separation for the two rules $R_1$ and $R_2$ by the following calculation for an arbitrary point $X$ of the cuboid:

$$\sum_{i=1}^{n} \mu_{R_2}(X) + 1 - n = \sum_{i=1}^{n} \mu_{R_1}(X) + 1 - n$$

$$\Leftrightarrow \sum_{i=1}^{n}(1 - (1-\alpha) \cdot \tfrac{x_i - a_i}{b_i - a_i}) + 1 - n$$

$$= \sum_{i=1}^{n}(1 + (1-\alpha) \cdot \tfrac{x_i - b_i}{b_i - c_i}) + 1 - n$$

$$\Leftrightarrow \qquad -\sum_{i=1}^{n}(1-\alpha) \cdot \tfrac{x_i - a_i}{b_i - a_i} = \sum_{i=1}^{n}(1-\alpha) \cdot \tfrac{x_i - b_i}{b_i - a_i}$$

$$\Leftrightarrow \qquad \sum_{i=1}^{n}(-x_i + a_i + b_i - x_i) = 0$$

$$\Leftrightarrow \qquad \sum_{i=1}^{n}(a_i + b_i) - 2 \cdot \sum_{i=1}^{n}(x_i) = 0 \qquad (5.3)$$

The last equation is exactly the right side of equation (5.2) multiplied by $-2\sqrt{n}$. Geometrically equation (5.3) is the hyperplane in the middle between the two opposite corners $(a_1, \ldots, a_n)$ and $(b_1, \ldots, b_n)$ orthogonal to the line between $(a_1, \ldots, a_n)$ and $(b_1, \ldots, b_n)$. The middle point $M$ belongs to this hyperplane. On one side of the calculated hyperplane the points are $\mathcal{C}_1$-possible and on the other side for $\mathcal{C}_2$.

This way of defining the chosen fuzzy sets solves our problem of the unclassified region inside the cube, but now the fuzzy sets are continued outside the cube as shown in figure 5.3 with the dashed line. With increasing degree of overlap the number of fuzzy sets that happen to have a positive membership degree inside the cuboid also increases (compare [34]).

It occurs that the rules do not stop firing outside the cube to that they belong. To solve this problem, we assume that the fuzzy sets are not continued for $x_i < a_i$ or $x_i > b_i$ if $[a_1, b_1] \times \cdots \times [a_n, b_n]$ is the cube that we consider (see [35] for other possible partitions).

To outline how the fuzzy system has to look like, we assume that the given fuzzy classification system consists of fuzzy sets of the form

$$\mu_{R_j}^{(i)}(x_i) = \begin{cases} 1 - \gamma_i \cdot |x_i - c_k^{(i)}| & \text{for } |x_i - c_k^{(i)}| \le \tfrac{1-\alpha}{\gamma_i} \\ 0 & \text{for } |x_i - c_k^{(i)}| > \tfrac{1-\alpha}{\gamma_i} \end{cases} \qquad (5.4)$$

being uniformly distributed in each dimension. $\gamma_i$ is the same for all fuzzy sets for the $i^{th}$ coordinate and $c_k^{(i)}$ is the middle of each fuzzy set. For $\delta_c^{(i)}$ and for two such points $c_{k_1}^{(i)}$ and $c_{k_2}^{(i)}$ belonging to neighboring fuzzy sets the equation $|c_{k_1}^{(i)} - c_{k_2}^{(i)}| = \delta_c^{(i)}$ holds. The resulting fuzzy sets look as in figure 5.6.

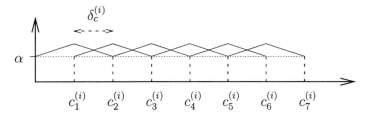

Figure 5.6: The fuzzy sets of the characterized fuzzy classification system

**Remark 14**

*We deal with classification instead of function approximation. The results of the two systems have different intentions, although both systems use rules with overlapping fuzzy sets.*

*Approximation normally aims at reproducing continuous functions. Therefore it is necessary to take into account also small membership degrees of the fuzzy sets and the small firing degrees of the resulting rules. These firing degrees contribute to obtain continuous functions and smooth transitions between the rules.*

*Considering a classification problem as we do here requires a discrete output, because the system has to come to a decision between the possible classes also at those points where the firing degrees for the different classes are nearly equal. There may occur two different cases when determining the classification of a point $P$:*

*In the first case there are rules firing with relatively high firing degree in $P$. Then other rules with small firing degrees can not change the result that is determined by the rules with greater firing degree. Finally they are of no consequence.*

*In the second case there are only rules firing in $P$ with a small firing degree. There is a classification resulting from these rules, but because of the small firing degrees it is not a clear classification. As we demand a strict distinction between the classes, it is sensible not to consider these small firing degrees.*

By cutting off the fuzzy sets at the sides we take all the small and therefore not very important firing degrees out of considerations. The system as shown in figure 5.6 can also be interpreted as a system with fuzzy sets that are meeting at $\alpha = 0.5$ but that are raised above the dotted line.

When we have a fuzzy classification set as the one defined here, then it can be characterized by constructing the separation. In the following section we will describe how to do this. For calculating the separating hyperplanes we only have to consider one cube at a time, as the fuzzy sets do not adopt values greater that 0 outside the cube.

Figure 5.7: Impossible and possible joining of two fuzzy sets at a corner

**Remark 15**
Let $[a_1, b_1] \times \cdots \times [a_n, b_n]$ be the cube. Figure 5.7 shows the situation at the right side of the interval $[a_i, b_i]$. The fuzzy sets inside the cube reaches membership degree 1 in $b_i$. The rule that can be found for $x_i > b_i$ prolongates the fuzzy sets for $x_i < b_i$ as to be seen in the right drawing. As the fuzzy set for $x_i > b_i$ also reaches membership degree 1 in $b_i$, the case of the left drawing can not appear.
Therefore we can consider the fuzzy set for $x_i > b_i$ as a prolongation of that one for $x_i < b_i$.

## 5.3 Several Rules inside a Single Cuboid

In the previous section we discussed single rules. The only case with two rules that we examined was the one with two opposite rules. Such a combination of two rules does not result in the rules influencing each other because the sections where they fire are disjoint.
In this section we analyze the behavior of several rules when they interfere with each other. This means that they fire in the same section of the cube. Then it is necessary to determine which rule has the greatest firing degree. First we consider two rules, then we explain how a greater number of rules can be examined.
We still assume, that $(b_i - a_i)$ is equal for all $i \in \{1, \ldots, n\}$, so that we have a cube for the cuboid.

We start by examining the case of two arbitrary rules firing for different classes and examine what sort of separation can result from different combinations of the rules. Then it will be easier to understand more complicated constructions.
As described in the proof of corollary 4, all the rules start with firing degree 0 at the middle of the cube, i.e. at a hyperplane passing by $M = (\frac{1}{2}(a_1 + b_1), \ldots, \frac{1}{2}(a_n + b_n))$ and increase orthogonally until they reach the corner where $\mu_{R_i}^{(i)} = 1$ for all $i \in \{1, \ldots, n\}$. Now we have to examine how these

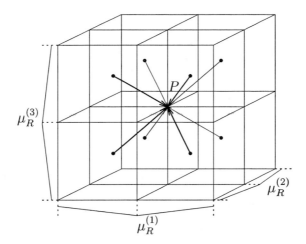

Figure 5.8: All the parts of one rule increase towards the same corner

rules interfere with each other.

As there are $2^n$ corners, we can have at most $2^n$ rules. Each rule can be associated with one corner, but not necessarily each corner represents a rule. The rule $R$, that is increasing from the middle $M_j$ of a cube $C_j$ towards the corner $P$, is also increasing from the middle $M_k$ of all those neighboring cubes $C_k$, that share the corner $P$ with $C_j$, towards $P$. This can be seen in figure 5.8. For better visibility we just marked the direction in which the firing degree is increasing and left out the starting planes.

The corner $P$ that represents a rule $R$ has the coordinates $p_i$ where the fuzzy sets yield $\mu_R^{(i)}(p_i) = 1$. These fuzzy sets increase from inside of the cube and decrease into the neighboring cube. Analyzing the adjacent cube, we realize that the extensions of the fuzzy sets into this cube also compose a rule that starts at the middle point of the adjacent cube and increases until it yields 1 in the corner $(p_1, \ldots, p_n)$.

Figure 5.8 shows how the different parts of the same rule all increase from the middle points of the eight neighboring cuboids to the same corner. As the same rule is coming from all directions towards one corner, we do not need any separating hyperplanes following the boundaries of the cube in this point.

Now we examine the separation resulting from two rules $R_1$ and $R_2$ firing for $\mathcal{C}_1$ and $\mathcal{C}_2$ in one cube. Let the fuzzy sets for $k$ coordinates be the same for the two rules and different for the other $n - k$ coordinates. Without loss of generality let $\mu_{R_1}^{(i)}(x_i) = \mu_{R_2}^{(i)}(x_i)$ for $i = 1, \ldots, k$ and $\mu_{R_2}^{(i)}(x_i) = 1 + \alpha - \mu_{R_1}^{(i)}(x_i)$ for $i = k+1, \ldots, n$ with $\mu_{R_1}^{(i)}(x_i) = 1 - \frac{1-\alpha}{b_i - a_i}(x_i - a_i)$. Then the two points

$P_1$ and $P_2$ belonging to the rules $R_1$ and $R_2$ have $k$ fuzzy sets in common, and differ by the fuzzy sets for $n - k$ coordinates, i.e. $x_{k+1}, \ldots, x_n$. We calculate the separating hyperspace by

$$
\begin{aligned}
\sum_{i=1}^{n} \mu_{R_1}^{(i)}(x_i) + 1 - n &= \sum_{i=1}^{n} \mu_{R_2}^{(i)}(x_i) + 1 - n \\
\Leftrightarrow \qquad \sum_{i=k+1}^{n} \mu_{R_1}^{(i)}(x_i) &= \sum_{i=k+1}^{n} (1 + \alpha - \mu_{R_1}^{(i)}(x_i)) \\
\Leftrightarrow \qquad 2 \sum_{i=k+1}^{n} \mu^{(i)}(x_i) &= (n-k)(1+\alpha) \\
\Leftrightarrow \quad \sum_{i=k+1}^{n} (1 - \frac{1-\alpha}{b_i - a_i}(x_i - a_i)) &= \frac{(n-k)(1+\alpha)}{2} \\
\Leftrightarrow \qquad -(1-\alpha) \sum_{i=k+1}^{n} \frac{x_i - a_i}{b_i - a_i} &= -(1-\alpha)\frac{n-k}{2} \\
\Leftrightarrow \qquad \sum_{i=k+1}^{n} \frac{x_i - a_i}{b_i - a_i} &= \frac{n-k}{2}
\end{aligned}
$$

(5.5)

The resulting space is an (n-1)-dimensional subspace and represents the space, where the two rules have the same firing degree. We call it a *separating hyperspace of degree $n - k$*, because there are $n - k$ fuzzy sets different for the two rules that result in this separation. Its normal vector is $n_S = (0, \ldots, 0, \frac{1}{b_{k+1} - a_{k+1}}, \ldots, \frac{1}{b_n - a_n})$.

**Corollary 5**
*This hyperplane is situated in the middle between the two corners of the cube in that the rules reach firing degree 1. It is orthogonal towards the straight line between these two corners.*

**Proof:** In the case we considered in (5.5) the two corners are the points $(b_1, \ldots, b_k, b_{k+1}, \ldots, b_n)$ for $R_1$ and $(b_1, \ldots, b_k, a_{k+1}, \ldots, a_n)$ for $R_2$. Their difference is $d = (0, \ldots, 0, b_{k+1} - a_{k+1}, \ldots, b_n - a_n)$, and $n_S \| d$.  $\square$

It is important to mention that the separating hyperplane of degree $n$, that results from two rules that do not have a single fuzzy set in common, is exactly that plane, where the two rules start with membership degree 0.

## 5.4   Separation for the Whole Data Space

We have different separating hyperplanes that result from the various combinations of rules that fire in the cuboid. There are e.g. $n$ separating hyperspaces of degree 1 and $n * (n - 1)$ of degree 2. Altogether we have $\frac{n!}{k!}$ possible separating hyperplanes of degree $k$ inside one cube.
We have to take into account that the separations extend those of the neighboring cuboids. Figure 5.9 shows an example with the continuously drawn

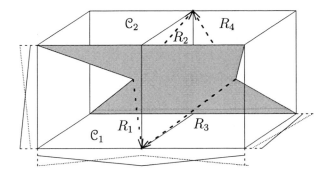

Figure 5.9: The separating hyperplanes can be extended into the neighboring cuboids.

fuzzy sets for the rules firing for $\mathcal{C}_1$ and the dotted ones for rules firing for $\mathcal{C}_2$. We can see, that the separating hyperplane resulting from $R_3$ and $R_4$ is the prolongation of that one resulting from $R_1$ and $R_2$.

To make figure 5.9 readable, we did not draw the hyperplanes, where the rules start, but they are neither firing in the whole cube nor inducing a separation in the whole cube. For that reason the separating hyperplane is not continued towards the boundaries of the cube. Either there are other rules than these four or part of the data in the cube is not classified.

As the mathematical equation describing a hyperplane is the same for a separating hyperplane as it is for its prolongation in the neighboring cuboid, we can combine them into one continuous linear separation. We call this the *extended separating hyperplane.* We use the following method to construct the separation:

For each cuboid we examine all the possible combinations of two rules that are firing inside the cuboid for different classes. For each combination of two rules we get a separating hyperplane. We use this separating hyperplane without considering where it begins or ends. In this way we get a number of sections that are defined by the extended separating hyperplanes. Because of not considering beginnings or endings of the hyperplanes, we diminish the number of separating hyperplanes, although there are more sections defined by the total number of extended separating hyperplanes than the hyperplanes inside the cuboids define.

To compute the class a section belongs to, we simply choose a point that is situated inside the section and calculate the class it belongs to. Then the whole section belongs to this class.

Until here we did not consider the boundaries of the region, where the separation hyperplanes are relevant. Therefore, there can be found neighboring sections that can be combined into one section.

For each section and for each separating hyperplane we have to know, on which side of the hyperplane the section is situated. We can simplify the description of the sections: If there are e.g. two sections described by "above $H_1$, above $H_2$ and below $H_3$" and "above $H_1$, below $H_2$ and below $H_3$", then we can join them into one single section described by "above $H_1$ and below $H_3$". These simplifications will be described more detailed in chapter 9.

Because there may appear a number of separating hyperplanes (e.g. $H_4$, $H_5$ and $H_6$) that are parallel, it is sufficient to know, between which two of the parallel hyperplanes the section can be found: If e.g. a point is above $H_4$ and below $H_5$, then it has to be below $H_6$.

## 5.5   Consequences

We have demonstrated an innovative way of characterizing a Łukasiewicz fuzzy classification system. This enables us to better understand the Łukasiewicz fuzzy classification system.

It is possible to draw a similar characterization for a Łukasiewicz fuzzy clustering system. When both systems can be characterized in a comparable way, then we are capable to derive a fuzzy classification system out of the fuzzy clustering system. We will explain this method in section 7.

We have seen the correlation between a fuzzy classification system and a geometric characterization of classified data by the means of separating hyperplanes. One benefit from these results is that the fuzzy classification systems can be characterized, so that humans can easier understand them, but in the following chapters we will demonstrate another benefit concerning fuzzy clusters.

# Chapter 6

# Fuzzy Clustering

Fuzzy clustering (for an overview see for example [37, 41]) is a technique to join with similar attributes in one group. Finding groups with comparable attributes in a data set leads to further understanding of the structure of the data.

As our input data consists of real-valued tuples or vectors with $n$ components, we assume that our data lie within $\mathbb{R}^n$ or - in case of normalized data - within $[0, 1]^n$. We usually assume that the data lie within some box $X_1 \times \ldots \times X_n \subseteq \mathbb{R}^n$, where the sets $X_i$ are intervals. To each datum one of the classes $\mathcal{C}_1, \ldots, \mathcal{C}_c$ or *unknown* is assigned. Including the class *unknown* into our classification means that we do not require that each datum must be classified in terms of the meaningful classes $\mathcal{C}_1, \ldots, \mathcal{C}_c$. This means that the subsets of $X_1 \times \ldots \times X_n$ associated with the classes (including *unknown*) induce a partition of $X_1 \times \ldots \times X_n$.

This partition or, equivalently, the assignment of the elements of $\mathbb{R}^n$ or $X_1 \times \ldots \times X_n \subset \mathbb{R}^n$ to the classes must either be specified by some expert or has to be learned from a finite training data set for which the classes are known.

## 6.1 Fuzzy Classification Systems

This section provides a brief introduction into the background of fuzzy classifiers and fuzzy clustering that is needed for understanding the connection that we are going to establish in chapter 7.

### 6.1.1 Fuzzy Classifiers

We remind some facts to make it easier to point out the differences between fuzzy classifiers and fuzzy clustering:

A fuzzy max-min classifier is characterized by a finite set $\mathcal{R}$ of rules of the form

$$R: \text{If } x_1 \text{ is } \mu_R^{(1)} \text{ and } \ldots \text{ and } x_n \text{ is } \mu_R^{(n)} \text{ then class is } \mathcal{C}^{(R)},$$

where $\mathcal{C}^{(R)}$ is one of the classes $\mathcal{C}_1, \ldots, \mathcal{C}_c$. The $\mu_R^{(i)}$ are assumed to be fuzzy sets on the intervals $X_i$, i.e. $\mu_R^{(i)} : X_i \to [0,1]$, where $X_i$ is an interval. In order to keep the notation simple, we denote the fuzzy sets $\mu_R^{(i)}$ directly in the rules. In real systems one would replace them by suitable linguistic values like *positive big*, *approximately zero*, etc. and associate the linguistic value with the corresponding fuzzy set.

A single rule is evaluated by interpreting the conjunction in terms of the minimum, i.e.

$$\mu_R(p_1, \ldots, p_n) \;=\; \min_{i \in \{1,\ldots,n\}} \left\{ \mu_R^{(i)}(p_i) \right\} \tag{6.1}$$

is the degree to which rule $R$ fires.

$$\mu_{\mathcal{C}}^{(\mathcal{R})}(p_1, \ldots, p_n) \;=\; \max \left\{ \mu_R(p_1, \ldots, p_n) \mid \mathcal{C}^{(R)} = \mathcal{C} \right\} \tag{6.2}$$

is the degree to which the point $(p_1, \ldots, p_n)$ is assigned to class $\mathcal{C}$.

Finally, the data point $(p_1, \ldots, p_n)$ has to be assigned to a unique class (defuzzification) by

$$\mathcal{R}(p_1, \ldots, p_n) \;=\; \begin{cases} \mathcal{C} & \text{if for all } \mathcal{C}' \neq \mathcal{C}: \\ & \mu_{\mathcal{C}}^{(\mathcal{R})}(p_1, \ldots, p_n) > \mu_{\mathcal{C}'}^{(\mathcal{R})}(p_1, \ldots, p_n) \\ unknown & \text{otherwise.} \end{cases}$$

This means that we assign the point $(p_1, \ldots, p_n)$ to the class of the rule with the maximum firing degree. If there is more than one class having the maximum firing degree, the class of the datum is *unknown*. Note that $\mathcal{R}$ denotes the set of rules as well as the associated mapping assigning to each point the corresponding class based on these classification rules.

We can easily generalize fuzzy max-min classifiers to fuzzy *s-t* classifiers where $t$ is a an arbitrary t-norm and $s$ a t-conorm by replacing the minimum in (6.1) by $t$ and the maximum in (6.2) by $s$.

An early overview on fuzzy classification is given in [68].

Although fuzzy classifiers were used in practical applications for many years already, studies about their fundamental properties have been published only in recent years. In [46] it was shown that max-min classifiers cannot solve linear separable classification problems for $n > 2$ exactly, but by the use of other t-norms or t-conorms than min resp. max, any linear separable problem can be solved by a fuzzy classifier. In [94] it was proved that max-min classification depends locally on only two variables.

Nürnberger et al. [79, 80] investigated the class boundaries of two- and three-dimensional data that can be generated by fuzzy classifiers using different t-norms. Cordón et al. [17] analyze fuzzy classifiers on an experimental basis that do not rely on a classification based on the rule that best fits the input. L. Kuncheva provides a very thorough and detailed analysis of fuzzy classifiers in [57].

It is important to note that in case of max-min classifiers as well as for other types of classifiers based on the Łukasiewicz t-norm or the bounded sum t-conorm the separation boundaries between classes can be described by hyperplanes, if triangular or trapezoidal membership functions are used in the rules. [13, 11] The dimension of these separating hyperplanes is always $n-1$ if $n$ attributes are used for the classification.

## 6.1.2 Fuzzy Clustering

Cluster analysis is used to identify sets with similar data. Data with comparable attributes are merged into the same cluster, while data with different attributes should be found in different clusters.

The resulting clusters of fuzzy clustering can be used to classify the data by assigning one cluster to one class. Note that it makes sense to join several clusters into one class.

We restrict our considerations to the most basic fuzzy clustering techniques: The fuzzy c-means algorithm with its variants. Other clustering techniques that also concider class information while clustering can be found e.g. in [52, 104].

The purpose of the fuzzy c-means algorithm [7] is to (fuzzy) partition a finite data set $\{x_1, \ldots, x_N\} \subseteq \mathbb{R}^n$ into a fixed number $c$ of clusters. Each cluster is represented by a prototype $v_i \in \mathbb{R}^n$ and for each datum $x_j$ we have to determine a membership degree $u_{ij} \in [0,1]$ to the cluster $v_i$. The prototypes and membership degrees should be chosen in such a way that they minimize the sum of the weighted distances of the data to the prototypes, weighted with the corresponding membership degrees, i.e.

$$\sum_{i=1}^{C} \sum_{j=1}^{N} u_{ij}^m d_{ij} \tag{6.3}$$

should be minimized where $d_{ij} = \| v_i - x_j \|^2$ is the squared Euclidean distance between prototype $v_i$ and datum $x_j$.

The parameter $m > 1$ is called fuzzifier and controls how well-separated fuzzy clusters are. It is not of importance in our context here.

In order to avoid the trivial solution $u_{ij} = 0$ for all $i, j$ – no datum is assigned to any cluster – an additional constraint has to be introduced:

$$\sum_{i=1}^{C} u_{ij} = 1 \qquad \text{for all } j. \tag{6.4}$$

This means that each datum is required to have an accumulated membership degree of 1 to all clusters. Since in this case the membership degree can be interpreted as the probability that a datum is assigned to a cluster, this kind of clustering is also called probabilistic clustering.

In order to have a minimum of the objective function (6.3) satisfying constraint (6.4) the following necessary conditions must be satisfied:

$$u_{ij} = \frac{1}{\sum_{k=1}^{C} \left(\frac{d_{ij}}{d_{kj}}\right)^{\frac{1}{m-1}}} \tag{6.5}$$

$$v_i = \frac{\sum_{j=1}^{N} u_{ij}^m x_j}{\sum_{j=1}^{N} u_{ij}^m} \tag{6.6}$$

In order to minimize the objective function (6.3) these two equations are alternatingly applied until convergence is reached. (Note that in the case $d_{ij} = 0$ the membership degree $u_{ij}$ must be set to 1 for exactly one $i$ for which $d_{ij} = 0$ holds.)

For a new datum $x \in \mathbb{R}^n$ that was not involved in the computation of the prototypes and the membership degrees, we can still determine its membership degree to any cluster $v_i$, simply by replacing $x_j$ in (6.5) by $x$. In this way we can associate a membership function $u_i : \mathbb{R}^n \to [0, 1]$ to each cluster $v_i$.

Due to the probabilistic constraint, the membership functions $u_i$ have some undesired properties. Since the membership degree in (6.5) depends only on the relative distance of the datum to the prototypes, the membership degree tends to $1/C$ for data that are far away from any cluster. Figure 6.1 illustrates this behavior by considering one-dimensional data and two prototypes at $v_1 = 0$ and $v_2 = 1$. The membership function shown is the one associated with cluster $v_1$.

In order to reduce this undesired effect, Davé [18] has introduced noise clustering that still uses the probabilistic constraint (6.4). However, in noise clustering there is one so-called noise cluster that is supposed to take care of

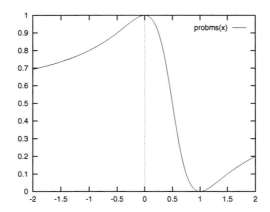

Figure 6.1: Probabilistic membership function

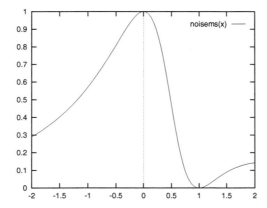

Figure 6.2: Noise membership function

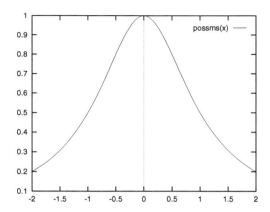

Figure 6.3: Possibilistic membership function

outliers. The noise cluster is not represented by a prototype. Instead it is assumed that each datum has a fixed (large) distance to the noise cluster.
Figure 6.2 shows the same membership function as in figure 6.1, except that we have introduced a third cluster, the noise cluster to which all data have a constant distance of 2.
Possibilistic clustering [53] drops the probabilistic constraint (6.4) completely and introduces an additional term into the objective function (6.3) assigning a penalty to membership degrees near 0.
This leads to the formula

$$u_{ij} = \frac{1}{1 + \left(\frac{d_{ij}}{\eta_i}\right)^{\frac{1}{m-1}}}$$

for the membership degrees. $\eta_i$ is a parameter determining how slow membership degrees decrease with increasing distance from the prototype. The equation for the prototypes remains the same as (6.3).
Looking at the right ends of the graphs in figures 6.1,6.2 and 6.3 we can see how outliers are treated by the corresponding fuzzy clustering strategies. Although possibilistic clustering seems to be the best choice for handling outliers, there are other disadvantages of possibilistic clustering as they are discussed in [19].
Although it is assumed that the number of clusters $K$ is fixed in advance, there are techniques to automatically adapt the number of clusters. For an overview on this topic we refer to [37].
Fuzzy clustering is designed for unsupervised classification, i.e., the assignment of the training data to classes is not known in advance. Nevertheless,

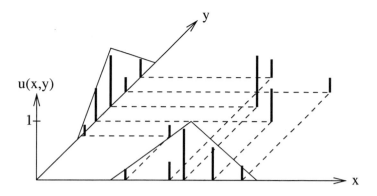

Figure 6.4: A projection of a fuzzy cluster

fuzzy clustering can also be applied in the case of supervised classification in order to construct a fuzzy classifier. One can for instance cluster the data first ignoring the class information and then assign a class to each cluster, namely the class to which the majority (in terms of the sum of the membership degrees) of the data assigned to this cluster belongs [31, 47]. An alternative approach for (partially) classified data is proposed in [83].

A fuzzy classifier based on fuzzy clustering can use the multi-dimensional membership functions $u_i$ directly. However, in order to better understand and interpret the classifier, it is desirable to describe the fuzzy classifier in terms of fuzzy rules as they were introduced in the previous subsection. Projecting the fuzzy clusters and approximating the projections by triangular or trapezoidal fuzzy sets is a very common strategy to derive rules from fuzzy clusters [47, 100]. In this way each cluster induces a rule using the fuzzy sets derived from the projections and the class associated with the cluster. Figure 6.4 illustrates the principle idea behind this concept.

Unfortunately, the projection and the approximation of the projections by standard membership functions enforces a certain loss of information, so that the classifier based on the rules does not always classify the data in the same way as it is done by the original multi-dimensional membership functions. When we look at the final classification after defuzzification, the only important parameters for the classification are the distances of the datum to be classified to the prototypes, no matter whether we apply probabilistic, noise, or possibilistic clustering. A datum is assigned to the class of the nearest prototype. Therefore, the crisp partition induced by the prototypes has hyperplanes as boundaries between the sets of the partition.

These hyperplanes can be used to construct a fuzzy classifier that produces

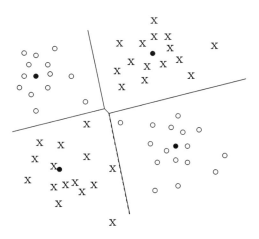

Figure 6.5: Cluster boundaries

exactly the same class boundaries. In [95] a rough idea of how a fuzzy classifier can be constructed on the basis of hyperplanes is outlined. Since in [94] it is proven that a max-min classifier decides locally on the basis of two attributes, such a classifier can only model hyperplanes that are parallel to $n - 2$ coordinates. To get more flexibility, we use the Łukasiewicz-t-norm instead of the minimum.

Note that the Łukasiewicz-t-norm is nilpotent, i.e. for all $\alpha \in (0, 1)$ there is an $n \in N$ with $\underbrace{(\alpha * \cdots * \alpha)}_{n \text{ times}} = 0$. Therefore it is not only 0 if at least one of its arguments is 0. Therefore fuzzy sets in a rule base that is valuated by the Łukasiewicz-t-norm should be chosen wider than in the case of the minimum. For a comparison of the Łukasiewicz-t-norm with the min-t-norm see figures 2.7 and 2.8.

## 6.2 Transforming Cluster Information into Hyperplanes and Cuboids

Assume that we have a given set of hyperplanes that result from the prototypes of the clusters. Each cluster is assigned to one class. For calculating these hyperplanes we take those hyperplanes that are situated exactly between two prototypes and that have the normalvector in the direction of the connection between the prototypes. In the worst case this would give us $\binom{c}{2}$ hyperplanes if we have $c$ clusters, but if two clusters belong to the same class

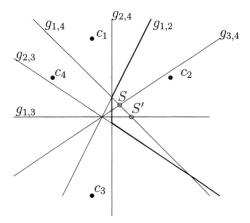

Figure 6.6: Four prototypes and the resulting separation lines

then we can remove the separating hyperplane between them.

Now we evaluate all those intersection points of $n$ hyperplanes that can be found inside the space $X_1 \times \ldots \times X_n$. Not all these points are relevant. It can happen that one of the hyperplanes belonging to an intersection point $S$ is not relevant at this point.

**Definition 10**

*A separating hyperplane $H$ is called irrelevant in $S \in \mathbb{R}^n$ if there is an $\epsilon \in \mathbb{R}^+$ so that in a neighborhood $\mathcal{N}_\epsilon(S)$ of $S$ the separation into classes does not depend on $H$, i.e.*

$$H \text{ irrelevant in } S \in \mathbb{R}^n \Leftrightarrow \exists\, \epsilon > 0 \ \forall\, h \in H \cap \mathcal{N}_\epsilon(S) \backslash S \ \exists \text{ class } \mathcal{C} \ni \mathcal{N}_\delta(h) \subset \mathcal{C}.$$

**Example 8**

*In figure 6.6 an example is shown. The $c_i$ are the prototypes of four clusters, and the $g_{i,j}$ are the separation lines that result from two clusters $c_i$ and $c_j$. The section bounded by $g_{1,2}$, $g_{2,4}$ and $g_{2,3}$ contains all those points that belong to the cluster with the prototype $c_2$.*

*In this example, the intersection point $S$ of $g_{1,4}$ and $g_{3,4}$ is irrelevant: As it is situated on $g_{1,4}$ and $g_{3,4}$, its distance to $c_1$, $c_3$ and $c_4$ is the same for all three clusters, but $g_{1,2}$ and $g_{2,4}$ decide that the distance to $c_2$ is less that to the other three clusters. Therefore it is unimportant, whether the membership degree is greater for $c_1$ or for $c_4$.*

*The same reflections can be made for the intersection point of $g_{1,3}$ and $g_{1,4}$.*

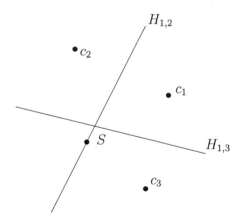

Figure 6.7: The point $S$ is irrelevant

In the intersection point $S$, the distance towards the cluster centers to that the intersecting hyperplanes belong, is the same. If there is another hyperplane that is not passing by $S$, but that gives a decision, then we can remove $S$. We will illustrate this with figure 6.7.

Let $H_{1,2}$ be a hyperplane passing by $S$ that is distinguishing between the classes $C_1$ and $C_2$. On this hyperplane the membership degree for both of them is equal. There is another hyperplane $H_{2,3}$ that distinguishes between $C_1$ and $C_3$ not passing by $S$. If $S$ is on that side of the second hyperplane opposite to $C_1$, then we can remove $S$ from our list, because $S$ does not belong to the cluster of $C_1$ anyway. Obviously this can only be the case if $S$ is not the intersection point of $H_{1,2}$ and $H_{1,3}$. Therefore, we do not need $H_{1,2}$ on this side of $H_{1,3}$.

When we remove the superfluous hyperplane, then there is no intersection point to be considered anymore.

After having examined all the intersection points like this, a list of relevant intersection points $S = \{S_1, \ldots, S_r\}$ remains and these points are the basis to construct a grid that divides our space into several cuboids, and for each cuboid, we can use the previously given construction for the fuzzy rules. The corners of the data space should be joined to $S$.

First of all we define for each $i \in \{1, \ldots, n\}$ the set

$$K_i := \{\alpha | \exists j \in \{1, \ldots, r\} \text{ so that } \alpha \text{ is the } i^{th} \text{ component of } S_j\}$$

of possible grid point values for the $i^{th}$ coordinate. Note that the sizes $k_i = |K_i|$ of these sets can be different for each coordinate, because it may happen that two points of $S$ share a coordinate.

When taking all combinations of these components we get at most $n^r$ points that span a grid in the space. This gives us up to $(n+1)^r$ cuboids.

**Remark 16**
*If there is no point $S_j \in S$ on the boundary between two neighbored cuboids, then we can put them together to form one common cuboid.*

By this we can diminish the number of cuboids step by step until all neighboring cuboids have a point $S_j \in S$ on their common boundary. Then we can take each single cuboid and construct the fuzzy rules for the separation done inside this cuboid.

For using the following algorithm it makes sense to sort the elements of the $K_i$ in increasing order in a vector. Then these vectors can be combined into an array of vectors $K$. A vector $k$ contains the sizes $k_i$ of the $K_i$.

Note that the function *Initialize* constructs a list of all combinations of the elements of the $K_i$. The resulting points form a grid that includes all the intersection points of the separating hyperplanes.

We choose the first cuboid of the list in a way that makes sure that we first consider all the cuboids with the lowest values for $x_i$.

Further explanations can be found in the appendix B.

**Algorithm 2 (Evaluation of the Cuboids)**

```
VectorList Initialize(VectorArray K)
{
    for all (x₁, ..., xₙ) ∈ {K₁[1], ..., K₁[k₁ − 1]} × {Kₙ[1], ..., Kₙ[kₙ − 1]}
        H.AddVector(x₁, ..., xₙ);                    //represents a cuboid
    return H;
}
```

```
SearchFurther(dimension i, Vector Start, Vector End )
{
    int s; Endᵢ := Endᵢ + 1;
    Startᵢ := Endᵢ;
    if (S ∩ Cuboid(Start, End)) = 0)
    then
        SearchFurther(i, Start, End);
    else
        if (S ∩ (Cuboid(Start, End) without its corners) = 0)
        then
            end of joined cuboid reached in direction xᵢ;
```

```
        else
                already too far, one step back;
        return s;
}

RectangleList Devision_Of_Dataspace(SetOfPoints S, K_1,...,K_n)
{
    H = Initialize(K);
    RectangleList List = empty list of rectangles;

    Cuboid Start:=First_Cuboid(H);
    while( H ≠ EMPTY )
        End:= Start; for all i = 1,...,n
            End_i := SearchFurther(i, Start, End);
        Remove Cuboids from Start to End from H;
        Add Cuboid(Start, End) to List;
    return List;
}
```

◇

## 6.3   Complementary Remarks

Now we have given a geometric visualization of the classification done by the fuzzy clusters. As our aim is to construct a fuzzy classification system out of this information, our results are only useful with a discribtion of the construction of such a fuzzy classification system. The following chapter will demonstrate this construction.

We will consider each cuboid individually. As the fuzzy sets will have a positive membership degree only inside one cuboid, we can put the resulting rules together and obtain a rule basis that classifies the data space and reproduces exactly the classification given by the fuzzy clusters.

The following chapter will describe in detail how to derive the fuzzy classification rules from clusters using the Lulasiewicz-t-norm.

# Chapter 7

# Fuzzy Classification Rules from Fuzzy Clusters

Fuzzy clustering (for an overview see for example [37]) is a method for joining data with comparable attributes into groups. Usually fuzzy clustering is applied in the context of unsupervised classification, where the data from the training set are not assigned to classes. But there are also methods to use fuzzy clustering in the case of supervised classification. Since a classifier derived from fuzzy clustering uses multi-dimensional membership functions, the corresponding classifier is often transformed into a fuzzy classifier using if-then rules in order to have a better interpretation and understanding of the classifier.

Projection is a very common technique to derive rules from fuzzy clusters. However, projection means always a loss of information so that the rule-based classifier does not have the same performance as the original one. This means that a certain degradation of accuracy is tolerated for the sake of interpretability.

The properties of fuzzy classifiers based on if-then rules are well-examined, and it is important to note that the structure of the class boundaries for standard fuzzy if-then classifiers and for fuzzy clustering classifiers is identical. We therefore propose not to derive rules by projecting fuzzy clusters, but to construct a fuzzy if-then classifier directly from the class boundaries induced by fuzzy clustering with fuzzy rules that reflect exactly these boundaries.

For this purpose we just assume that the classification problem is piecewise linearly separable. We use the Łukasiewicz-t-norm, as it does not restrict the resulting fuzzy classification system but allows to map any piecewise linear separation and therefore any separating hyperplane between two clusters. We can construct a fuzzy classification system based on this norm, that draws exactly these boundaries [97].

In chapter 4, we have demonstrated how to construct a fuzzy classifier for a linearly separable problem in the two-dimensional case. Here we want to do the same with the n-dimensional case. After explaining the basic principle on the case of three dimensions, we explain the calculation for a subcuboid of the space and finally specify how to handle the whole data space. In the end of this chapter we will demonstrate the method with the Iris data set.

# 7.1   Łukasiewicz Classification System with h Hyperplanes

The geometric characterization of the Łukasiewicz-t-norm was described in 4.2. In the n-dimensional space such a rule starts with firing degree 0 at a hyperplane and then increases orthogonally to this hyperplane until it reaches the firing degree 1. When we have two different classes on both sides of a hyperplane $H$, then we just need two rules, one for the first class that starts with 0 at $H$ and increases into the direction of the class that it represents, and the second rule to do the same into the other direction.

In this case at each time we have one rule with a firing degree greater than 0, but when there are several hyperplanes describing the classes, then we have to use several rules firing for one datum. This is to be described in the following.

We assume that we have only two classes. To get the more general case, we just have to consider whether a datum is in a class $C$ or not. In the following step this can be done one after the other for the remaining classes to find the correct classification.

In chapter 6, we have already seen that the classification based on fuzzy clustering is characterized by hyperplanes. We want to construct a classifier that defines exactly the same hyperplanes as class boundaries.

As a first step we do not consider the whole data space but partition it into cuboids of the form $[a_1, b_1] \times \ldots \times [a_n, b_n]$. The boundaries of the cuboids form a lattice.

The cuboids are chosen in such a way that they contain $h$ hyperplanes that intersect in one point and define a convex region. Inside this region we have one class, outside this region another class (compare figure 7.4). The point of intersection should be placed in one corner of the cuboid, but can also be situated outside.

**Remark 17**

*Because the intersection points have to be in the corners of the cuboids, it is reasonable to use the intersection points for the construction of the lattice.*

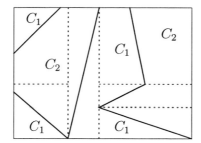

Figure 7.1: An example how to partition the space into handy rectangles

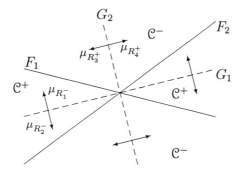

Figure 7.2: The case of type two

Figure 7.1 illustrates the partition of the space for a two-dimensional example. We will treat the partitioning process more detailed in section 7.2.

In order to be able to illustrate the construction of the rules, we will explain the method for the 3-dimensional case, but the technique easily extends to higher dimensions.

## 7.1.1 Three-Dimensions Examples

If we consider the classification of a three dimensional space, we try to use the same principle as for the two dimensional case. This means that we divide the space into cuboids that include as few planes as possible. We obtain rectangles of type 0 (no plane inside) to 3 (three planes meeting in one corner of the cuboid).

Type 0 is trivial, and type 1 can be solved analogously to the two-dimensional case in chapter 4 by separating into "above $\mathcal{D}$" and "below $\mathcal{D}$". with $\mathcal{D}$ being a separating hyperplane.

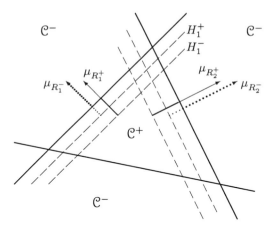

Figure 7.3: In a cuboid of type three

In a rectangle of type 2, two planes $F_1$ and $F_2$ meet. In Figure 7.2, the projection plane is located perpendicular to the two planes that meet. We add two "auxiliary planes" $G_1$ and $G_2$, that are located exactly between $F_1$ and $F_2$. Now the rules have to be constructed that way that they start with membership degree 0 at $G_1$ resp. $G_2$ and then increase perpendicularly as shown in figure 7.2. By varying the slope we can adjust them to meet with equal membership degree at $F_1$ resp. $F_2$.

When the top section e.g. is to belong to $\mathcal{C}^+$ instead of $\mathcal{C}^-$, then we can change $\mu_{R_1^-}$ into $\mu_{R_1^+}$ or simply leave it out.

The case of a cuboid of type 3 as shown in figure 7.4 becomes more complicated. Here we have to use a t-conorm. We choose the maximum as t-conorm. Figure 7.3 shows a plane that intersects the three separating planes.

For each separating plane we add two imaginary planes. For the $i^{th}$ separation plane, they are called $B_i^+$ and $M_i^-$. Now we define for each plane a rule $\mu_{R_{B_i}}$ resp. $\mu_{R_{M_i}}$ that starts with membership degree 0 at this plane and increases outside the triangle. At the separation plane they both have the same membership degree.

To make sure that no part of the inner section remains unclassified, the auxiliary planes $M_i^-$ should all meet in one line inside the section. This line starts in the point $P_s$ where the three separating planes meet and passes by an arbitrary point inside the section, for instance the center of gravity.

We get in our example a correct classification for $\mathcal{C}^+$ inside and for $\mathcal{C}^-$ outside the three planes.

For the three- and higher dimensional case, we need a t-conorm and the structure of the rules becomes more and more complex. This will be described mode detailed in the following and can be generalized to arbitrary dimensions.

## 7.1.2  Basic Principles

As we suppose to have a linearly separable classification problem, it is possible to partition the space into several $n$-dimensional cuboids that have hyperplanes for the separation that have only few breaks inside the cuboid. In the 2-dimensional case, we have to divide the cuboid again into several smaller ones at the breaking points as illustrated in figure 7.1. The higher-dimensional case is more challenging.

A break in a separating hyperplane can be considered as intersection of two straight hyperplanes. On each side of the intersection one of the two planes in considered. In the $n$-dimensional space the critical points are the intersections of $n$ hyperplanes, having the dimension 0. We use these points to divide the cuboids. This procedure provides us with a set of cuboids. None of the cuboids has an intersection point inside. A possible resulting cuboid is shown in figure 7.4.

In order to consider the region, that is described by the hyperplanes, as a bounded set, we sometimes have to consider the borders of the cuboid as additional separating hyperplanes. The number of hyperplanes defining the sector equals $n$.

Now we have a region that is bounded by the different hyperplanes. The interior of the region belongs to one class, while the outer part belongs to another class as shown in figure 7.4. The most interesting points for the construction of the fuzzy classifier are those intersection points that we also used for the devision of the cuboids.

The very special case with more than $n$ hyperplanes meeting in the same point will not be considered here, but the construction would follow the same principles.

We will depict the basic principles of the construction in the three-dimensional case. We have $n = 3$ hyperplanes that meet in one point $P_s$. They mark a section that belongs to one class, while the surroundings belong to another class (see figure 7.4). To construct the rules we use a straight line $l_M$ that starts at $P_s$ and is continued inside the section. We can draw it e.g. through the center of gravity of the section. Figure 7.5 shows a cut through the section that is orthogonal to this straight line.

Now we construct two auxiliary planes for each plane. These planes represent those points, where the assigned rules adopt the value 0. The first auxiliary plane $M_i$ includes $l_M$ and its image in figure 7.5 is parallel to $H_i$, while $B_i$

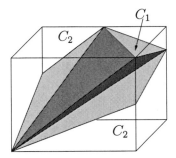

Figure 7.4: Three hyperplanes partition the space into two classes- inside and outside the section that is marked by the planes.

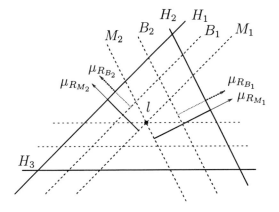

Figure 7.5: A cut through the three-dimensional case orthogonally to the straight line $l_M$.

is situated in the middle between $H_i$ and $M_i$. The rule $R_{B_i}$ that starts with firing degree 0 at $B_i$ increases twice as fast as $R_{M_i}$, so that it "overtakes" $R_{M_i}$ exactly at $H_i$.

The resulting system consists of two rules for each hyperplane. One rule $R_{M_i}$ gives a firing degree for the class inside the section. The other rule $R_{B_i}$ has a firing degree lower than that one of $R_{M_i}$ inside the section, but at $H_i$ it adopts the same firing degree, and outside the section it is the winning rule itself.

**Remark 18**

*If there are less than $n$ separating hyperplanes, so that the sector continues up to the boundaries of the section, then we do not use the center of gravity to construct $l_M$. Instead we use a point on the involved boundaries of the cuboid.*

*In this case we do not need to construct rules for these boundaries, as the class is the same on both sides of the boundaries.*

When we construct these two rules for each hyperplane, we get a fuzzy classification system that solves our classification problem correctly. The next section describes the calculations for the construction of the rules in the $n$-dimensional case.

## 7.1.3   Steps for the Construction of the Classification System

Let $P_s$ be the point where all the hyperplanes $H_1, \ldots, H_h$ meet. We can assume that $h \leq n$. We want to distinguish between the section that is bounded by the hyperplanes and the region outside this section. Two hyperplanes meet in one "line" (hyperplane of dimension $n - 2$). These lines of intersection between two hyperplanes are called $l_{ij}$, $i, j \in \{1, \ldots, h\}$, $i \neq j$. As we just consider one section marked by the hyperplanes, there are only $\lambda$ lines that are relevant. Those $l_{ij}$, that are irrelevant, are situated outside the inner section and therefore do not belong to it's boundaries. Each hyperplane is only involved in the definition of two lines, those where it meets its neighbor.

First of all we have to calculate the center of gravity of the marked section. We need the vectors $x_k$, $k = 1, \ldots, \lambda$, that are directed from $P_s$ to those points, where the $l_{ij}$ meet the border of the rectangle. Then the center of gravity $P_g$ is calculated by

$$P_g := P_s + \frac{1}{\lambda} \cdot \sum_{i=1}^{h} x_i.$$

By drawing a line between $P_s$ and $P_g$ we get a line

$$l_M : x = P_s + \alpha(P_g - P_s), \qquad (\alpha \in \mathbb{R})$$

that is situated inside the section that we want to describe by the fuzzy classification system and that passes by $P_s$.

## Algorithm 3 (Initialization)

*Initialize(Cuboid $[a_1, b_1] \times \cdots \times [a_n, b_n]$, HyperplaneList H)*
*{*
    *$P_s$ := intersection point of $H_1, \ldots, H_h$;*
    *ListofHyperlines $l$ := list of intersection lines $l_{ij} := H[i] \cap H[j]$*
        *for all $i, j = 1, \ldots, h$;*
    *double $\lambda$ := number of hyperlines $l_{ij}$;*
    *VectorArray X;*
    *length of $X$ := $\lambda$;*
    *for all $i = 1, \ldots, \lambda$*
        *$X[i]$ := $\lambda \cap$ Boundaries of the cuboid;*
    *$P_g$ := $P_s + \frac{1}{\lambda} \cdot \sum_{i=1}^{\lambda} x_i$ = center of gravity;*
    *Line $l$ :  $x = P_s + \alpha(P_g - P_s)$;*
*}*

$\diamond$

### Construction of the Auxiliary Planes

Now we have to construct the auxiliary planes. The following construction has to be done for each hyperplane $H_i$ separately.

First we need the vector $y_{H_i}$ that belongs to $H_i$, and that can be written as a linear combination of $n_{H_i}$ and $(P_g - P_s)$ with $n_{H_i}$ being the normal vector of $H_i$:

$$y_{H_i} = \alpha \cdot n_{H_i} + \beta \cdot (P_g - P_s), \qquad (\alpha, \beta \in \mathbb{R}).$$

The principle can be seen in figure 7.6, that shows a cut through $l_M$ and $H_i$. This cut includes $l_M$ and is perpendicular to $H_i$. If $\alpha > 0$, then $n_{H_i}$ is pointing into the direction of $l_M$, otherwise into the other direction.

Now we construct an orthogonal basis for $H_i$ that includes $y_{H_i}$. Replacing in this basis $y_{H_i}$ by $(P_g - P_s)$, we get a basis for the auxiliary plane $M_i$. We calculate the normal form $0 = n_{M_i} \cdot x + d_{M_i}$ of $M_i$ with the normal vector $n_{M_i}$. When choosing a point $P_{H_i}$ of $H_i$, we calculate $p_{H_i} = n_{M_i} \cdot P_{H_i} + d_{M_i}$.

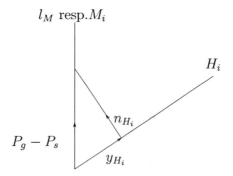

Figure 7.6: $y_{H_i}$ is a linear combination of $(P_g - P_s)$ and $n_{H_i}$.

If $p_{H_i} > 0$, then $n_{M_i}$ is pointing into the direction of $H_i$, otherwise into the other direction. The same can be done for a point $P_{M_i}$ of $M_i$.

For $B_i$ has to be in the middle between $H_i$ and $M_i$, we use the normal vectors $n_{B_i}$ and $n_{H_i}$ of the two planes. The normal vector $n_{B_i}$ of $B_i$ is calculated by

$$n_{B_i} := \frac{n'_{M_i} - n'_{H_i}}{|\, n'_{M_i} - n'_{H_i}\, |}$$

if $n'_{H_i}$ is pointing towards $M_i$ and $n'_{M_i}$ towards $H_i$. This can be achieved by using $n'_{M_i} = \text{sgn}(p_{H_i}) \cdot n_{M_i}$ and $n'_{H_i} = \text{sgn}(p_{M_i}) \cdot n_{H_i}$.

As $P_s$ has to belong to $B_i$, we can calculate $d_{B_i} = -P_s \cdot n_{B_i}$. Then $B_i$ is described by

$$n_{B_i} \cdot x + d_{B_i} = 0.$$

The exact formula for the construction can be found in appendix B.2 on page 165 .

**Algorithm 4 (Construction of the Auxiliary Planes)**

Auxiliary_Planes(Hyperplane $H$, $P_g$, $P_s$)
{
    Calculate a $y_{H_i} \in H_i$ with $y_{H_i} = \alpha \cdot n_{H_i} + \beta \cdot (P_g - P_s)$, $\alpha, \beta \in \mathbb{R}$;
    Calculate an orthogonal basis of $H_i$ including $y_{H_i}$;
    Basis of $M_i$ := Basis of $H_i$ with $y_{H_i}$ changed into $(P_g - P_s)$;
    Check direction of the normalvectors;
    Calculate $n_{B_i}$ and $d_{B_i}$;
    return $(M, B)$;
}

$\diamond$

**Construction of the Rule $\mathbf{R_{M_i}}$**

Now we have to determine the rules $R_{M_i}$ and $R_{B_i}$, that belong to the two planes $M_i$ and $B_i$. The firing degrees of the rules have to start with $\mu = 0$ at $M_i$ resp. $B_i$.

As the t-norm is the Łukasiewicz t-norm, we calculate $R_{M_i}$ and $R_{B_i}$ by

$$R_{M_i}(x_1, \ldots, x_n) = \sum_{t=1}^{n} \mu_{R_{M_i}}^{(t)}(x_t) + 1 - n \text{ and}$$

$$R_{B_i}(x_1, \ldots, x_n) = \sum_{t=1}^{n} \nu_{R_{B_i}}^{(t)}(x_t) + 1 - n.$$

First of all we construct the rule $R_{M_i}$ that has to start with $R_{M_i}(X) = 0$ at any $X \in M_i$ and to increase until it reaches $R_{M_i}(P_i) = 1$ at the corner $P_i = (p_1, \ldots, p_n)$ of the cuboid.

The membership degrees have to be between 0 and 1, therefore all fuzzy degrees have to be 1 in $P_i$ to fulfill $\sum_{t=1}^{n} \mu_{R_{M_i}}^{(t)}(p_t) + 1 - n = 1$. As we want to have linear fuzzy sets, we choose

$$\mu_{R_{M_i}}^{(t)}(x_t) = \begin{cases} 1 - \alpha_t \cdot (x_t - p_t) \\ 1 - \alpha_t \cdot (p_t - x_t) \end{cases} \text{ if } \begin{array}{l} p_t = a_t \\ p_t = b_t \end{array} \tag{7.1}$$

with $[a_1; b_i] \times \ldots \times [a_n; b_n]$ being the cuboid. Then the $\alpha_t, t = 1, \ldots, n$, are the unknown values of the fuzzy sets and have to stay between 0 and 1. We can denote this by

$$\mu_{R_{M_i}}^{(t)}(x_t) = 1 + \alpha_t' \cdot (x_t - p_t) \tag{7.2}$$

with $\alpha'_t = \alpha_t$ if $p_t = a_t$ and $\alpha'_t = -\alpha_t$ if $p_t = b_t$. Now we have to calculate the values $\alpha'_t$. Let the hyperplane $M_i$ be described by

$$\sum_{t=1}^{n} \gamma_t \cdot x_t + c = 0. \tag{7.3}$$

The values $\gamma_t$ are the components of the normal vector $n_{M_i}$ of $M_i$. We multiply the equation with $\gamma := \frac{-1}{c+\sum_{t=1}^{n} \gamma_t \cdot p_t}$, so that we get

$$\sum_{t=1}^{n} \gamma'_t \cdot x_t - \frac{c}{c + \sum_{t=1}^{n} \gamma_t \cdot p_t} = 0 \tag{7.4}$$

with $\gamma'_t = \gamma \cdot \gamma_t$ instead of equation (7.3). As the firing degree of the rule has to be 0 at the hyperplane $M_i$, this rule has to fulfill the condition

$$\sum_{t=1}^{n} \mu_{R_{M_i}}^{(t)}(x_t) + 1 - n = \sum_{t=1}^{n}(1 - \alpha'_t \cdot (x_t - p_t)) + 1 - n = -\sum_{t=1}^{n} \alpha'_t \cdot (x_t - p_t) + 1 = 0 \tag{7.5}$$

Therefore we define $\alpha'_t := \gamma'_t$. With this construction, the equations (7.4) and (7.5) are equivalent:

$$\mu_{R_{M_i}}(x_1, \ldots, x_n) = 0 \qquad \Leftrightarrow$$

$$\sum_{t=1}^{n} \alpha'_t \cdot x_t - 1 - \sum_{t=1}^{n} \alpha_t \cdot p_t \quad = \quad \gamma \sum_{t=1}^{n} \gamma_t \cdot x_t - 1 - \gamma \sum_{t=1}^{n} \gamma_t \cdot p_t \quad =$$

$$\gamma(\sum_{t=1}^{n} \gamma_t \cdot x_t - (-c - \sum_{t=1}^{n} \gamma_t \cdot p_t) - \sum_{t=1}^{n} \gamma_t \cdot p_t) \qquad =$$

$$\gamma(\sum_{t=1}^{n} \gamma_t \cdot x_t + c) \qquad\qquad = \quad 0$$

$$\Leftrightarrow \quad \sum_{t=1}^{n} \gamma_t \cdot x_t + c \qquad\qquad = 0$$

$$\Leftrightarrow \quad (x_1, \ldots, x_n) \in M_1,$$

so that the $\mu_{R_{M_i}}^{(t)}$ define a rule that starts at the hyperplane $M_i$ with firing degree 0 and increases until it reaches firing degree 1 at the point $P_i$.

The result is the same, if we have the the normal vector of $M_i$ pointing into the opposite direction, i.e. when we have $-n_{M_i}$ instead of $n_{M_i}$.

The fuzzy sets that we have constructed also take values that do not belong to $[0, 1]$. By scaling and cutting them in the very end we obtain fuzzy sets that range between 0 and 1. This will be described on page 99.

**Algorithm 5 (Construction of the Rule $R_{M_i}$)**

*void MRule( Hyperplane $M_i$, $H_i$, Cuboid $[a_1, b_1] \times \cdots \times [a_n, b_n]$,*
     *int class_inside, class_outside)*

```
{
    Choose corner Pᵢ that belongs to Hᵢ;
    γ := −(d_{M_i} + ∑ⁿ_{t=1}(n_{M_i}[t] · P[t]))⁻¹;
    for all j = 1, . . . , n
        α := γ · n_{M_i}[j]
        if (P[j] = a_j)
        then
            μ^(_{R_{M_i}}j)(xᵢ) := 1 − n_{M_i}[j] · (x_j − P[j]);
        else
            μ^(_{R_{M_i}}j)(x_j) := 1 − n_{M_i}[j] · (P[j] − x_j);
        Add_to_R (μ_{R_{M_i}}, class_inside);
}
```

                           $\diamond$

The detailed algorithm can be found in the appendix on page 166.

The corner that is used for the construction of the rules has to be the corner that is the farthest from the hyperplane and that is outside the section. The idea for choosing the corner is the following:

We start with a point inside the section. We can use the center of gravity $P - g$ for this purpose. Then we consider each coordinate an move towards one border value, i.e. either to $a_i$ or to $b_i$.

First we want to get out of the section. That means that we chose $a_i$ or $b_i$ depending on which one leads us further toward the hyperplane, or which one moves the point even to the other side of the hyperplane.

When the point is outside the section we have to take care that we do not cross the hyperplane again, and we choose the coordinate that leads us to the point with the greatest distance from $H$.

## Construction of the Rule $R_{B_i}$

Now we turn towards the other rule $R_{B_i}$ that has to start at the hyperplane $B_i$ and increase faster than $R_{M_i}$ until it "overtakes" $R_{M_i}$ at the hyperplane $H_i$. For the construction we first consider the fuzzy sets $\nu_i$ to be linear, i.e. we also allow them to adopt values below 0 or above 1. Later we cut them at 0 and 1 so that we get

$$\nu^{(t)}_{R_{M_i}}(x) = \max\{0, \min\{1, b_t + \beta_t \cdot x\}\}.$$

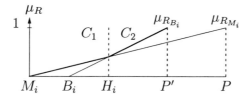

Figure 7.7: The rule for the class $C_2$ 'overtakes' the rule for $C_1$.

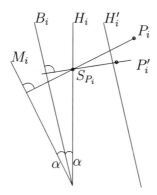

Figure 7.8: An orthogonal cut through the planes and $P_i$.

In figure 7.7 we illustrate how these rules behave. The horizontal line represents the way from $M_i$ to $P_i$ passing the other two auxiliary planes, while the vertical axis shows the firing degree of the rules ($R_{M_i}$ and $R_{B_i}$)

We can use the same construction as in the previous section, if we have a point $P_i'$ that fulfills the function that $P_i$ has for $R_{M_i}$, i.e. that the rule reaches firing degree 1 at $P_i'$. As $B_i$ is situated in the middle between $M_i$ and $H_i$, and as $R_{B_i}$ has to reach the same firing degree at $H_i$ as $R_{M_i}$, the rule $R_{B_i}$ has to increase twice as fast as $R_{M_i}$.

For the calculation of $P_i'$, we have to consider the construction as shown in figure 7.8. Let $P_i'$ lie in the plane that is orthogonal to $M_i$, $B_i$ and $H_i$ and that includes $P_i$. this is the planes that is shown in figure 7.8. Let $S_{P_i}$ be the point where $H_i$ and the line from $P_i$ to $M_i$, that is orthogonal to $M_i$, meet. The rules $R_{M_i}$ and $R_{B_i}$ are to have the same firing degree on $H_i$ and therefore also in $S_{P_i}$.

Now we have to put $P_i'$ that way that

$$\frac{\text{dist}(P_i, M_i)}{\text{dist}(S_{P_i}, M_i)} = \frac{\text{dist}(P_i', B_i)}{\text{dist}(S_{P_i}, B_i)}$$

with the distance $\text{dist}(P_i, H) = P_i \cdot n_{H_i} + d_H$ if a hyperplane $H$ is described by the normal form $P_i \cdot n_{H_i} + d_H = 0$.

$$\frac{\text{dist}(P_i, M_i)}{\text{dist}(S_{P_i}, M_i)} = \frac{\text{dist}(P'_i, B_i)}{\text{dist}(S_{P_i}, B_i)}$$

$$\Leftrightarrow \quad (S_{P_i} \cdot n_{B_i} + d_{B_i})(P_i \cdot n_{M_i} + d_{M_i}) = (S_{P_i} \cdot n_{M_i} + d_{M_i})(P'_i \cdot n_{B_i} + d_{B_i})$$

$$\Leftrightarrow \quad P'_i \cdot n_{B_i}(S_{P_i} \cdot n_{M_i} + d_{M_i}) = \frac{(S_{P_i} \cdot n_{B_i} + d_{B_i})(P_i \cdot n_{M_i} + d_{M_i})}{(S_{P_i} \cdot n_{M_i} + d_{M_i})} - d_{B_i}$$

$$(7.6)$$

As the right side of the last equation is a scalar, we get the normal form of a hyperplane $H'_i$ that is parallel to $B_i$. The firing degree of the rule $R_{B_i}$ is increasing orthogonally to $B_i$ until it reaches 1 at $H'_i$.

Now we can choose the point $P'_i \in H'_i$. $P_i$ has to belong to $H'_i$ and to be on the line $S'_{P_i} + \alpha \cdot n_{B_i}$, $\alpha \in \mathbb{R}$. When we construct the rule $R_{B_i}$ the same way as we have constructed $R_{M_i}$ in the previous section starting with firing degree 0 at $B_i$ and increasing until it reaches 1 at $P'_i$, then also

$$R_{M_i}(a_1, \ldots, a_n) = R_{B_i}(a_1, \ldots, a_n) \text{ for all } (a_1, \ldots, a_n) \in H_i$$

is fulfilled.

## Algorithm 6 (Construction of the Rule $R_{B_i}$)

```
void BRule()
{
    Point P':= Choose_P'(Point P, Hyperplane M,B,H, Cuboid);
    Rule R_B := Calculate_Rule(P', B, [a_1, b_1] × ··· × [a_n, b_n])
    Add_to_R (R_B, class_outside);
}


Point Choose_P'(Point P, Hyperplane M,B,H, Cuboid)
{
    Point S;
    Vector n_H, n_M, n_B;
    int d_H, d_M, d_B, β;
    S := P − n_H · (P·n_H+d_H)/(n_M·n_H);
    β := ((S·n_B+n_B)·(P·n_M+n_M))/(S·n_M+n_M) − d_B;
    P' := S + (n_B · (β − S · n_B));
    return (P');
}
```

◇

In the appendix on page 168 the detailed algorithm can be found.

**Scaling the Rules**

Now we have the two rules that we needed to describe the classification performed by the hyperplane $H_i$. We have to do this for all the hyperplanes $H_1, \ldots, H_h$ separately and then combine them by using the maximum as t-conorm. Now we have to make sure that the rules $R_{M_i}$ and $R_{M_j}$ (resp. $R_{B_i}$ and $R_{B_j}$) do not disturb each other when they collide at the $l_{ij}$. Therefore the rules for the two hyperplanes that meet at $l_{ij}$ must have the same firing degree at this hyperline. Anyway the two rules for one hyperplane $H_i$ have the same firing degree $R_{B_i}(a_1, \ldots, a_n) = R_{M_i}(a_1, \ldots, a_n)$ for each point $(a_1, \ldots, a_n)$ of $H_i$. Now we require the rules for the two hyperplanes $H_i$ and $H_j$ to have the same membership degree $R_{M_i}(a_1, \ldots, a_n) = R_{M_j}(a_1, \ldots, a_n)$ for each point $(a_1, \ldots, a_n)$ of $l_{ij}$. Then obviously we also have $R_{B_i}(a_1, \ldots, a_n) = R_{B_j}(a_1, \ldots, a_n)$. We can choose any point $(a_1, \ldots, a_n) \in l_{ij}$.

These equalities can be achieved by scaling the fuzzy sets for the rules. The aim is that a rule $R_{M_i}$ reaches firing degree $\delta_i$ instead of 1 in $P_i$ while the rule $R_{B_i}$ reaches $\delta_i$ in $P_i'$. $\delta_i$ has to be in $]0; 1]$ and $\max_{i \in \{1, \ldots, h\}}\{\delta_i\} = 1$.

If we just consider two hyperplanes $H_i$ and $H_j$ meeting at $l_{ij}$ with

$$s_{ij} := \frac{\mu_{R_{M_i}}(a_1, \ldots, a_n)}{\mu_{R_{M_j}}(a_1, \ldots, a_n)}$$

for any $(a_1, \ldots, a_n) \in l_{ij}$, then $s_{ij}$ would be the scaling multiplier. We determine the scaling multiplier for each $l_k$. By using the equation $s_{il} := s_{ij} \cdot s_{jl}$ we calculate the other scaling multipliers, so that we have one for each pair of hyperplanes. We determine $s := \max_{i,j}\{s_{ij}\} = s_{pq}$, and then the $p$ tells us the hyperplane $H_p$ that stays the same, while the other $H_j$, $j \neq p$, are to be scaled with $s_{pj} < 1$.

This means that they are to reach the firing degree $s_{pj}$ in $P_i$ instead of the firing degree 1. As changing the firing degree of a rule in a point results in a complex system of equations, the easiest way to achieve this is to do the same construction as we described it in the previous sections for a point $P_i$ (and resp. $P_i'$) for a new point $\bar{P}_i$ (resp. $\bar{P}_i'$), that is situated further away from $M_i$. Let $S_{M_i}$ be the orthogonal projection of $P_i$ on $M_i$. If $\tau$ is defined by $S_{M_i} + \tau \cdot n_{M_i} = P_i$, then we choose a point

$$\bar{P}_i = S_{M_i} + \frac{1}{s_{pj}} \cdot \tau \cdot n_{M_i} = P_i - (1 - \frac{1}{s_{pj}}) \cdot \tau n_{M_i}$$

as to be seen in figure 7.9 instead of $P_i$ to construct the rules.

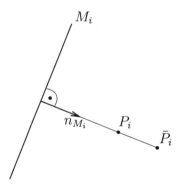

Figure 7.9: Rules belonging to $\bar{P}_i$ have less steep fuzzy sets.

If for the point $P_i$ the construction would result in a rule with the fuzzy sets $\mu_{R_{M_i}}^{(t)}(x_t) = 1 - \alpha_t(x_t - p_t)$, then the rule for $\bar{P}_i$ would result in

$$\bar{\mu}_{R_{M_i}}^{(t)}(x_t) = 1 - \alpha_t \cdot \frac{\bar{\gamma}}{\gamma} \cdot (x_t - p_t + \tau \cdot (1 - \frac{1}{s_{pj}}) \cdot n_{M_i}^{(t)}) \qquad (7.7)$$

with $\bar{\gamma} := -(c + \sum_{t=1}^n \gamma_t(p_t - \tau \cdot (1 - s_{pj}^{-1}) \cdot n_{M_i}^{(t)}))^{-1}$.   We calculate $\gamma = -(c + \sum_{t=1}^n \gamma_t \cdot p_t)$, and $\gamma_t = n_{M_i}^{(t)}$ being the $t^{th}$ coordinate of the normal vector $n_{M_i}$ of the hyperplane $M_i$ by analogy to page 95. The correctness of 7.7 can easily be shown by calculating

$$\mu_{\bar{R}_{M_i}}(x_1, \ldots, x_n) = \sum_{t=1}^n \bar{\mu}_{R_{M_i}}^{(t)}(x_t) + 1 - n$$

$$= n - \bar{\gamma} \sum_{t=1}^n \gamma_t(x_t - p_t + b \cdot n_{M_i}^{(t)}) + 1 - n$$

$$= 1 + \frac{\sum \gamma_t \cdot x_t - \sum \gamma_t \cdot (p_t - b \cdot n_{M_i}^{(t)})}{c + \sum \gamma_t \cdot (p_t - b \cdot n_{M_i}^{(t)})}$$

$$= 1 + \frac{c + \sum \gamma_t \cdot x_t}{c + \sum \gamma_t \cdot (p_t - b \cdot n_{M_i}^{(t)})}$$

with $b := \tau \cdot (1 - s_{pj}^{-1})$. The function $\mu_{\bar{R}_{M_i}}$ is linear. For all points $(a_1, \ldots, a_n) \in M_i$ we have $\sum_{t=1}^n \gamma_t \cdot a_t + c = 0$ and then we get

$$\mu_{\bar{R}_{M_i}}(a_1, \ldots, a_n) = \frac{0}{c + \sum \gamma_t \cdot (p_i - b \cdot n_{M_i}^{(t)})} = 0,$$

at $M_i$ and for the point $\bar{P}_i = P_i - b \cdot n_{M_i}$ we get

$$\mu_{\bar{R}_{M_i}} = 1 + \frac{c + \sum \gamma_t \cdot (p_t - b \cdot n_{M_i}^{(t)})}{c + \sum \gamma_t \cdot (p_i - b \cdot n_{M_i}^{(t)})} = 1.$$

When having done this with each rule the firing degree of two rules that meet at an $l_i$ is the same everywhere on $l_i$. This guarantees that those two rules do not disturb each other, when we choose the maximum-t-conorm, so that also the combination of the rules results in the given classification.

Then inside the section that is bordered by the $H_i$, $i = 1, \ldots, h$, we have the first class for that all the $R_{M_i}$ are firing and outside the section we have the other class.

The only thing still to be done is to make sure that the membership degrees of the fuzzy set stay in $[0, 1]$ by calculating

$$\mu_{\tilde{R}_{M_i}}^{(t)}(x_t) = \max\{0, \min\{1, \mu_{\tilde{R}_{M_i}}^{(t)}(x_t)\}\}$$

and resp. $\mu_{\tilde{R}_{B_i}}^{(t)}$. We do not get a change in the classification: By the time when the firing degree of a rule reaches 1, all the fuzzy sets have already membership degree 1, and from this on, we simply stay at this firing degree. And at the point, where a fuzzy set has membership degree 0, we have already a firing degree 0 for the rule, so that beyond it stays 0 anyway.

The algorithm can be found in the appendix on page 168.

## 7.2 Segmentation into Cuboids

Until now we just considered one cuboid of the dimension $n$ with hyperplanes of the dimension $n - 1$ that go from one side of the rectangle to the other side. The main point is that we have to classify the space by much more complicated separations, that can be combined of many hyperplanes as we have in the case of a piecewise linearly separable problem.

To use the construction that we developed for one cuboid, we have to partition the space into several cuboids that fulfill our conditions.

### 7.2.1 Regular Grid

We can cover the space with a regular grid. We mark those points where a hyperplane does not have to be continued, so that from this point on the separation follows a different hyperplane. Then we construct a grid, that is parallel to the coordinates and that includes all those points. An example can be seen in figure 7.10. In this example we need fifteen rectangles to cover the space.

This method has the advantage, that the resulting fuzzy sets can be transformed into uniformly distributed fuzzy sets by stretching the coordinates, but it can result in an awful lot of cuboids, and with a lot of cuboids we get many rules.

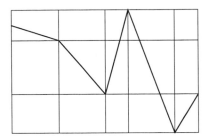

Figure 7.10: An example for covering the space with a regular grid.

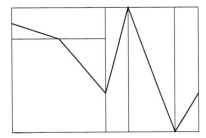

Figure 7.11: The same example when using a 'minimal' grid.

## 7.2.2   As few Cuboids as Possible

There is another possibility. Here we just construct a sort of minimal grid. Again we determine the points, where the directions of the separation planes change, but we just draw a part of the grid until we reach the next part of the grid resp. the next coordinate of one of the determined points. The algorithm for constructing this minimal grid is a separate problem and will not be discussed here.

When we consider the same example as in figure 7.10, then we just need five rectangles instead of fifteen as we can see in figure 7.11. The minimal grid is not unique.

This method using only few cuboids has the advantage of constructing only few rules. This makes it faster when calculating the rules themselves, but we need more time to determine the cuboids that are to be considered. Additionally the construction results into fuzzy rules that are not uniformly distributed on the coordinates.

**Remark 19**

*In section 7.3, we select the necessary hyperplanes heuristically. An algorithmic method will be given in section B.3. The principle of the algorithmic method is the following:*

1. *Check all possible intersection points of hyperplanes and boundaries of the data space, and decide whether they are relevant or not.*

2. *All clusters defining the intersection point via the hyperplanes have an equal distance to the intersection point.*

3. *An intersection point is relevant if it is located inside the data space and if there is no cluster that is closer to it than the defining clusters.*

4. *If an intersection point is relevant, all hyperplanes that define it, are relevant hyperplanes.*

## 7.3 An Example: The Iris Data Set

To illustrate our procedure, we use the well known iris data [24]. The data set contains three classes of 50 instances each. The classes are types of iris plants (iris setosa, iris versicolor, iris virginica). There are four input attributes, but clustering only with the last three attributes gives nearly as good results as using all four of them [42]. We refer to the used attributes as $x$, $y$ and $z$. The range of these attributes defines the cuboid $[2, 4.4] \times [1, 6.9] \times [0.1, 2.5]$ that is to be considered.

By using the fuzzy c-means algorithm we get five clusters as depicted in figure 7.12. The prototypes are the following.

$$c_1 = \begin{pmatrix} 3.43 \\ 1.46 \\ 0.25 \end{pmatrix}, c_2 = \begin{pmatrix} 2.52 \\ 3.82 \\ 1.15 \end{pmatrix}, c_3 = \begin{pmatrix} 2.93 \\ 4.53 \\ 1.43 \end{pmatrix}, c_4 = \begin{pmatrix} 2.89 \\ 5.20 \\ 1.93 \end{pmatrix}, c_5 = \begin{pmatrix} 3.13 \\ 5.99 \\ 2.16 \end{pmatrix}.$$

The cluster with the prototype $c_1$ belongs to class 0, $c_2$ and $c_3$ define clusters of class 1 and the clusters of $c_4$ and $c_5$ belong to class 3. Therefore we just have to consider the hyperplanes between $c_1$ and $c_2$ and between $c_3$ and $c_4$. These hyperplanes are

$$H_1 : \begin{pmatrix} -0.34 \\ 0.88 \\ 0.34 \end{pmatrix} \cdot x - 1.55 = 0 \text{ and } H_2 : \begin{pmatrix} -0.05 \\ 0.80 \\ 0.59 \end{pmatrix} \cdot x - 4.75 = 0$$

In figure 7.13 the prototypes of the clusters are depicted together with the separating hyperplanes between the first and second prototype and between the third and fourth.

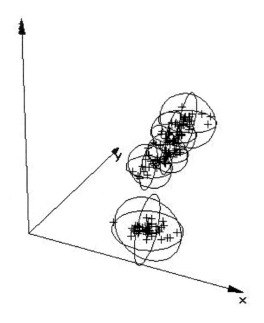

Figure 7.12: The Iris Data and its five clusters.

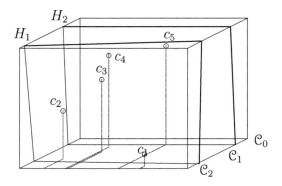

Figure 7.13: The prototypes of the clusters and the hyperplanes

We calculate the center of gravity $P_g$ out of the points, where $H_1$ and $H_2$ intersect the surface of the cuboid. $H_1$ and $H_2$ do not intersect inside the cuboid, and as we have only two hyperplanes, we use the y-z-boundary of the cuboid as a third hyperplane to determine the intersection point $P_s$.

$$P_g = \begin{pmatrix} 3.20 \\ 3.83 \\ 1.30 \end{pmatrix} \quad \text{and} \quad P_s = \begin{pmatrix} 2.0 \\ -1.304 \\ 9.938 \end{pmatrix}.$$

The next step is to calculate the auxiliary planes $M_1$ and $M_2$, and with their help $B_1$ and $B_2$:

$$M_1 : \begin{pmatrix} -0.39 \\ 0.81 \\ 0.43 \end{pmatrix} \cdot x - 2.42 = 0 \text{ and } M_2 : \begin{pmatrix} -0.07 \\ 0.86 \\ 0.50 \end{pmatrix} \cdot x - 3.72 = 0$$

$$B_1 : \begin{pmatrix} -0.37 \\ 0.85 \\ 0.38 \end{pmatrix} \cdot x - 1.98 = 0 \text{ and } B_2 : \begin{pmatrix} -0.06 \\ 0.83 \\ 0.55 \end{pmatrix} \cdot x - 4.24 = 0$$

It is obvious that the point $P_1 = (4.4,\ 1.0,\ 0.1)$ has to be chosen to calculate the rule $R_{M_1}$ and $P_2 = (2.0,\ 6.9,\ 2.5)$ to calculate $R_{M_2}$. The points $P_1' = (3.95,\ 1.83,\ 0.61)$ and $P_2' = (2.09,\ 5.93,\ 2.00)$ are closer towards the planes, so that the fuzzy sets of the rules $R_{B_1}$ and $R_{B_2}$ have a greater slope by absolute value than those of $R_{M_1}$ and $R_{M_2}$. $R_{B_1}$ and $R_{M_1}$ have the same firing degree at $H_1$ and $R_{B_2}$ and $R_{M_2}$ do so at $H_2$. We obtain the following rules:

Rule $R_{B_1}$:

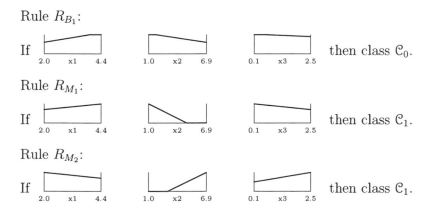

Rule $R_{M_1}$:

Rule $R_{M_2}$:

Rule $R_{B_2}$:

If 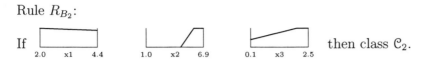 then class $\mathcal{C}_2$.

Now we have a rule base for a fuzzy classification system that classifies the iris data. The only misclassified data are those that have a greater distance to the prototype of their own class than to another prototype. This fuzzy classification system computes exactly the same classification as the fuzzy clustering does.

It can be seen from the fuzzy sets, that e.g. $R_{B_2}$ increases faster than $R_{M_2}$, and that $R_{B_1}$ and $R_{B_2}$ reach firing degree 1 before the corner of the data space.

## 7.4   Conclusions

Fuzzy clustering can be applied to unsupervised and even to supervised classification problems. However the description of classes in terms of cluster prototypes and multidimensional membership functions is not always suitable for interpreting the classification system. Therefore, a number of new techniques to derive fuzzy classification rules from the clusters have been proposed. These approaches usually project the clusters onto the single attributes and accept a loss of information leading to a less accurate classifier. In this chapter we have introduced a method that avoids projection and uses the class boundaries directly to derive the classification rules from the clusters. In this way an interpretable rule-based classifier is obtained that maintains the accuracy of the original clusters.

It should be emphasized that our method is restricted to fuzzy c-means clusters and cannot be extended to ellipsoidal clusters as they are computed e.g. in the Gustafson-Kessel algorithm [32],or to clusters of different sizes [43] since in this case the cluster boundaries are no longer hyperplanes.

# Chapter 8

# Perceptrons and Multilayer Perceptrons

In the precedent chapters we have used fuzzy rules to design classification systems. As we have seen in chapter 3, the common max-min inference leads to quite restricted classification systems that decide locally on the basis of only two variables ([94]). In order to build more flexible systems, we used the Łukasiewicz t-norm instead of the minimum. It was shown that such a fuzzy classification system (FCS) basically constructs a set of (hyper-)planes to separate the classes.

Since multilayer perceptrons rely in principle on the same strategy, our idea is to construct a fuzzy classifier on the basis of expert knowledge and then to transform it into a multilayer perceptrons in order to apply learning techniques to further reduce the classification error. The main advantage is that the neural network does not have to start learning from scratch, but begins already with a good initialization, assuming that the expert knowledge leads to a fuzzy system with an acceptable small error.

This chapter will introduce the perceptron leading to the description of the multilayer perceptron (MLP). Afterwards chapter 9 concentrates on the construction of a multilayer perceptron, based on a classification that uses hyperplanes for class separation.

Multilayer perceptrons as well as fuzzy classification systems assign at least approximately the input data to the classes by the means of hyperplanes. The difference is based on the fact that the MLP uses the means of the layers of the Neural Network to define the sectors, the FCS uses interference of the rules that are responsible for the boundaries of the sectors. Both systems have in common that they assign the data to the sectors and then to the classes.

Until now the thesis dealt with fuzzy classification systems. This chapter

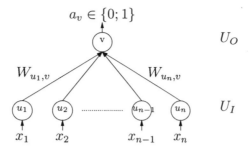

Figure 8.1: The perceptron for an n-dimensional input space

will be to explain the multilayer perceptron. As they both can be visualized as systems that draw separating hyperplanes and use them to assign sectors to classes, this method gives us a means to translate both systems into each other.

We first introduce the perceptron as a predecessor of the multilayer perceptron and than the more general case of a multilayer perceptron. In both cases it is important to know which sort of classifications problems can be solved.

## 8.1   The Perceptron

A perceptron is a very basic neural network. It is was introduced as extreme simplification of one cell in the human brain. A single unit is taking the input signals to evaluate a single output value.

A perceptron is a feed forward neural network with two layers. It has $n$ units in the input layer $U_I = \{u_1, \ldots, u_n\}$ and one unit in the output layer $U_O = \{v\}$. An activation function is assigned to each unit. For the input units $u_1, \ldots, u_n$ the activations $a_{u_1}, \ldots, a_{u_n}$ are determined by $a_{u_i} := x_i$ for $i \in \{1, \ldots, n\}$ with $X = (x_1, \ldots, x_n) \in \mathbb{R}^n$ being the input signal. The activation of the output unit $u_v$ is calculated by

$$a_v := \begin{cases} 0 & \text{if } \sum_{i=1}^{n} W(u_i, v) \cdot a_{u_i} - \Theta \leq 0, \\ 1 & \text{if } \sum_{i=1}^{n} W(u_i, v) \cdot a_{u_i} - \Theta > 0. \end{cases} \tag{8.1}$$

$\Theta$ is called *threshold value* or *bias*. The $W(u_i, v)$ are called *weights* of the connection from $u_i$ to $v$ and can be noted in a matrix $W : U_I \times U_O \to \mathbb{R}$. Figure 8.1 shows a perceptron for the n-dimensional input.

An example will visualize the functionality of the perceptron.

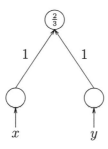

Figure 8.2: The perceptron that solves the AND-problem

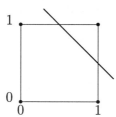

Figure 8.3: The AND solved by a perceptron

## Example 9 (The AND)

*We can use the perceptron to perform the logical AND. The perceptron shown in figure 8.2 consists of an input layer $U_I = \{u_x, u_y\}$ with two units for the two dimensional input space and one unit in the output layer $U_O = \{u_O\}$. The weights and the threshold value are the following: $W = (1, 1)$, $\theta = \frac{2}{3}$.*

*Figure 8.3 shows the space with the line that the perceptron uses to classify it. If the input values describe a point above the line, then the output is $a_{u_O} = 1$, otherwise 0.*

*The separating line is described by the function:*

$$f(x, y) = \sum_{x,y} W(u, v) \cdot a_u = 1 \cdot a_{u_x} + 1 \cdot a_{u_y}.$$

*If $f(x, y) = \Theta$, then the input point is exactly on the separation line. Equation 8.1 determines if the input points are above or below the separating line and results in the correct classification.*

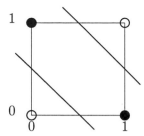

Figure 8.4: The input space and its separating lines

## 8.2   A Perceptron solves a Linearly Separable Problem

The logical AND is the most simple case in which the two classes can be distinguished by one hyperplane, this is the linearly separable case. A perceptron with $n$ input units is able to solve a linearly separable problem in an $n$-dimensional space.

We have two classes $\mathcal{C}^+$ and $\mathcal{C}^-$. Let a point belong to the class $\mathcal{C}^+$, when the output of the perceptron is 1, and to the complementary class $\mathcal{C}^-$, when the output is 0.

The separation between the two classes is defined by the hyperplane that is described by

$$\sum_{i=1}^{n} W(u_i, v) \cdot x_i - \Theta = 0.$$

Thus the perceptron can solve any linearly separable problem (see [76]).

The example 9 demonstrates that a perceptron can only solve linear separable problems. The logical XOR can not be solved by a perceptron, because it is not possible to separate the two classes by only one separation line. But we can combine several perceptrons.

**Example 10 (The XOR)**
*We consider the logical XOR that is obviously not linearly separable. But we can separate the points $(0,0)$ and $(1,1)$ from $(0,1)$ and $(1,0)$ if we use two separating lines as shown in figure 8.4.*

*Each line is represented by one unit of the first hidden layer, i.e. $U_1$. These two layers together are formed by $|U_1|$ perceptrons, each one consisting of the input layer and one unit of $U_1$.*

*The activations of the units of layer $U_1$ form the points shown in figure 8.5. As the input points $(0,1)$ and $(1,0)$ lie in the same section in figure 8.4, they*

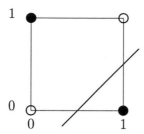

Figure 8.5: The values of the first hidden layer

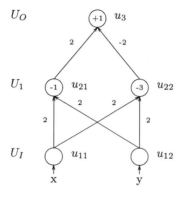

Figure 8.6: The resulting MLP to solve the logical XOR

*are mapped to the same point in figure 8.5.*

*There is one separating line needed to reach the final classification. This separating line can be realized by another perceptron that consists of $U_1$ as input layer and one output unit. The output unit indicates the classification. When we combine these three perceptrons, we can solve the XOR problem. The resulting neural network is shown in figure 8.6.*

*The activations of the units are calculated by the following:*

$$a_{u_{11}} = x \text{ and } a_{u_{12}} = y;$$
$$a_{u_{21}} = 2 \cdot a_{11} + 2 \cdot a_{12} - 1;$$
$$a_{u_{22}} = 2 \cdot a_{11} + 2 \cdot a_{12} - 3;$$
$$a_{u_3} = -2 \cdot a_{21} + 2 \cdot a_{22} + 1;$$

*The weights for one layer can be combined into one matric. The resulting weight matrices are*

$$W_{I1} = \begin{pmatrix} -2 & 2 \\ -2 & 2 \end{pmatrix} \text{ and } W_{1O} = \begin{pmatrix} -2 \\ 2 \end{pmatrix}.$$

## 8.3   The Multilayer Perceptron

In typical classification problems, the separation has to be much more complicated than a linear separation. Even if it is not possible to give a correct classification in any case, we can do an approximate separation by treating the problem as a piecewise linearly separable one. Our means for this is a multilayer perceptron (MLP). In the precedent example we have seen how a neural network can be built up from several perceptrons that are put in different layers one after another.

The multilayer perceptron (MLP) consists of $l$ layers $U_I = U_0, U_1, \ldots, U_{l-2}$, $U_O = U_{l-1}$. As the problem is an $n$-dimensional one, we have $n$ input units $u_1, \ldots, u_n$, and in the other layers $U_i$, $i \in \{1, \ldots, m-1\}$, we have $m_i$ units $v_{i_1}, \ldots, v_{i_{m_i}}$ with $W_{v_j} : U_i \times \{v_j\} \to \mathbb{R}$. Combining all the weights of the $v_j$, we achieve $W_{U_{i+1}} : U_i \times U_{i+1} \to \mathbb{R}$ for the weights from the layer $U_i$ to layer $U_{i+1}$.

The activations are calculated as in equation 8.1. This means that we consider the special case of a threshold activation function, but in chapter 9.5 we deal with the differentiable activation function.

The threshold function results from the construction of the MLP out of several perceptrons, but as there was no learning algorithm known, the search for a learning algorithm led to a sigmoidal activation function [90]. This means that the activation of a unit is calculated by the following equation:

$$a_v := net_v(u_1, \ldots, u_n) = f(\sum_{i=1}^{n} W(u_i, v) \cdot a_{u_i} - \Theta)$$

with $f : \mathbb{R} \to [0,1]$ resp. $f : \mathbb{R} \to [-1,1]$ being a sigmoidal function called *activation function*. One possibility for a sigmoidal function is the logistical function

$$f : \mathbb{R} \to [0,1], f(x) = \frac{1}{e^{-x}}.$$

**Remark 20**

1. *Each unit $v_j \in U_1$ represents one separation plane in the space by giving the output 1, if the point is on one side of this hyperplane, and 0 if the point is on the other side. When we have $m$ units in $U_1$, then we have $m$ hyperplanes that divide the space into up to $2^m$ segments. Assigning these segments to the classes gives the solution for piecewise linearly separable problems.*

2. *Using a sigmoidal function instead of a threshold function leads to a fuzzy separation. The more the output tends towards 0 and 1, the clearer is the separation. If the output is close to 0.5, then the input point is not clearly separated from the foreign class.*

Now we can combine several layers $U_0 = U_I, U_1, \ldots, U_{l-1} = U_O$ one after another in the same way and obtain a multilayer perceptron with $l$ layers. The input values are the activation values of the units in layer $U_k$ and the output value is the activation of the chosen unit in layer $U_{k+1}$. All activations adopt either value 0 or value 1. This means, that the weights of one layer $U_k$, $k \in \{1, \ldots, l-2\}$, to one unit $v_j$ of the following layer $U_{k+1}$ represent a boolean expression

$$\beta_{kj} : \{0; 1\}^m \rightarrow \{0; 1\}$$

with $m := |U_k|$. Putting one layer after another, we apply one boolean function after another. The result is a boolean function, that represents the assignment of the sections to the classes:

$$\text{sort} : \{0; 1\}^m \rightarrow \{0; 1\}^c$$

with $c := |U_o|$. We consider one unit of the output layer. This unit represents one class versus the union of the other classes. Like this we always have to consider only two classes. The activation of this unit is calculated out of the activations of the units of layer 2, that characterize the sections. If the unit's activation value is 1, then the input point belongs to the class represented by the unit, otherwise it does not belong to this class.

All points that belong to the same sector, have the same output and therefore belong to the same class. All the sectors that produce the same output form one class.

Although multilayer perceptrons are not capable of solving arbitrary problems, but they can approximate the separation by a linearly separable problem. This fact leads us to the parallels between multilayer perceptrons and fuzzy classification systems. In the next chapter, we will see how to construct an MLP if the hyperplanes and the classes of the sections are known.

# Chapter 9

# Construction of a Multilayer Perceptron for a Piecewise Linearly Separable Classification Problem

We consider a continuous space that is divided into several sections by hyperplanes. The sectors between these separation planes are assigned to classes, so that all the points of one section belong to the same class, while a class can consist of several sections.

We are interested in constructing a system, that determines the corresponding class of a given point. The problem that has to be solved is a piecewise linearly separable one. One possibility to solve this classification problem is a fuzzy classification system using the Lukasiewicz-t-norm. In chapter 7 (see [95]) we demonstrated, how to construct such a fuzzy classification system, when the separating hyperplanes are known.

Here we want to do the same with a multilayer perceptron. Neural networks, e.g. the multilayer perceptron, can be used for such a classification, if the output values are discrete. They learn their weights from given examples by supervised learning (see e.g. [77, 96]). The hyperplanes give us a connection between the fuzzy classification system and the multilayer perceptron.

A multilayer perceptron calculates for each input data the class it belongs to. For this purpose it draws hyperplanes through the data space and assigns the resulting sectors to the classes. For each input data the MLP decides to which sector it belongs and then assigns it to the associated class.

**Remark 21**
*This geometrical visualization only works for a multilayer perceptron. It does not hold for any other type of neural networks.*

We want to introduce an explicit construction of a multilayer perceptron that has already the correct weights and bias to solve the classification problem. This is useful e.g. if we want to initialize a neural network with approximately known separation planes and then improve the classification by learning.

We minimize the number of used hyperplanes to keep the size of the multilayer perceptron as small as possible. The big advantage of our method towards a brute force approach, that devides the regions of a class by auxiliary hyperplanes into simplexes and assigns them to classes, is the fact, that the resulting network is smaller.

In the classical definition of the multilayer perceptron, the activation of the units and with it the output space are continuous. The sigmoidal activation function enables the MLP to learn via backpropagation algorithm. When we want to use the multilayer perceptron for classification, then there is no sense to interpolate between the classes by using values between 0 and 1. When we define the sigmoidal activation function very flat, then the learning performs fast. When the activation function is very steep, i.e. nearly a threshold function, then the learning is slow but the classification is clear.

While considering the construction of the MLP, we restrict the output to 0 and 1. When performing the neural network later, we turn the activation function into a sigmoidal one.

When we have one output unit, the multilayer perceptron can distinguish between two classes. The unit gives 1 as output, if the classified point belongs to the class $\mathcal{C}^+$, and giving 0 if it does not belong to the class, which means that it belongs to the complementary class $\mathcal{C}^- = \mathbb{R}^n \backslash \mathcal{C}^+$.

When we have more than two classes, then we use one output unit for each class $\mathcal{C}_i \in \{\mathcal{C}_1, \ldots, \mathcal{C}_c\}$. The unit $w_i$ that represents the class $\mathcal{C}_i$ has the output $a_{w_i} = 1$, if the point belongs to this class, while the other output units all give 0.

In section 8.2 we have seen how a perceptron solves a linearly separable problem, and in section 8.3 we have explained the extension of the perceptron into a multilayer perceptron, that can solve piecewise linearly separable problems. Such a multilayer perceptron divides the classes by hyperplanes, so that we get several sectors. Each unit of the first hidden layer represents one of these hyperplanes. The units of the second hidden layer can be interpreted as the sections. Then the multilayer perceptron defines, which sector belongs to which class. In section 9.2 we explain the construction of such a multilayer perceptron if the hyperplanes are known.

The assigning of the sections to the classes can be represented by a boolean expression, and we consider the special cases of the conjunctive and the disjunctive normal form as they are described in the sections 9.3 and 9.4. If we know the hyperplanes and the boolean expression, then we can construct a multilayer perceptron that divides the classes exactly by the known hyperplanes.

In section 9.5 we see, that taking the classical definition for a multilayer perceptron that uses a sigmoidal activation function, we get sort of a membership degree for each class, so that we can see, if the classification is very clear, or if there is only a small distance towards another class.

## 9.1 Suppositions

Assume that a partition of the space $\mathbb{R}^n$ into $c$ classes $\mathcal{C}_1, \ldots, \mathcal{C}_c$ is given. If the neural network is to learn the classification, then we need a set of examples for supervised learning, and this supervised learning is to give us a classification function

$$\text{class} : \mathbb{R}^n \to \{\mathcal{C}_1, \ldots, \mathcal{C}_c\}.$$

Here we assume that the classification is already known, this means that there are hyperplanes $H_j \in \mathcal{H}$, $j \in \{1, \ldots, J\}$ given by the equation

$$\sum_{k=1}^{n} \alpha_k^{(j)} \cdot x_k - \Theta^{(j)} = 0.$$

These hyperplanes divide the space into several sections. Each section is characterized by the hyperplanes, more precisely we need to know for each point $(x_i, \ldots, x_n)$ of the section and for each hyperplane

$$\begin{array}{ll} \text{whether} & \sum_{k=1}^{n} \alpha_k^{(j)} \cdot x_k - \Theta^{(j)} \leq 0 \\ \text{or} & \sum_{k=1}^{n} \alpha_k^{(j)} \cdot x_k - \Theta^{(j)} > 0. \end{array}$$

For each section we have to know, to which class it belongs. This can be given e.g. by points with their assigned classes. For each section we need one point, that belongs to this section, with its class.

It can happen that two neighboring sections belong to the same class, so that the hyperplane separating them is redundant. E.g. in figure 9.1 on page 120, the sections "above $g_1$, below $g_2$, below $g_3$ and below $g_4$" and "below $g_1$, below $g_2$, below $g_3$ and below $g_4$" both belong to $\mathcal{C}^-$, so that we can combine them into the section "below $g_2$, below $g_3$ and below $g_4$".

We will realize the function "class" by a multilayer perceptron with $n$ input units for the $n$ dimensions of the data space and with $c$ output units for the $c$ classes using two hidden layers.

Here we just consider two classes $\mathcal{C}^+$ and $\mathcal{C}^-$, but the results easily extend to arbitrary numbers of classes.

## 9.2 Construction of a Multilayer Perceptron from Given Hyperplanes

Assume that we have a space that is divided by several hyperplanes into different sections, and each section belongs to one of the classes $\mathcal{C}^+$ and $\mathcal{C}^-$ (in the two-dimensional case e.g. as shown in figure 9.1).

There we have e.g. a rule

> If (x,y) is above $g_1$, above $g_2$, below or above $g_3$
> and below or above $g_4$,
>
> then it belongs to $\mathcal{C}^-$.

Such rules can be expressed e.g. by a fuzzy classification system that uses the Łukasiewicz-t-norm.

We assume that the hyperplanes are known and that we know to which class each section belongs. The latter can be described e.g. by a set of points. We need at least one point out of each section and the class this point belongs to.

Such sections can also be described by a multilayer perceptron with four layers. Figure 9.2 shows such an MLP for the example of figure 9.1. There we have two input units, because the space is two-dimensional. The four hidden units of layer $U_1$ belong to the four straight lines that divide the space into sections, and in layer $U_2$ the units represent the sections as we will see later. The output unit gives 1, if the input point belongs to the class $\mathcal{C}^+$, and 0 otherwise.

### 9.2.1 The logical AND

The logical AND is used to interpret the transformation from layer $U_1$ to $U_2$. Each unit of layer $U_1$ represents one hyperplane and describes whether the input point is above or below this hyperplane. The equation for the separating hyperplane represented by $v \in U_2$ is given by $\mathrm{net}_v$.

By combining these pieces of information by the AND, we get the characterizations of the sectors. E.g. "$a_1$ AND NOT $a_2$ AND $a_3$" characterizes the sector that is above $P_1$, below $P_2$ and above $P_3$, while "$a_1$ AND $a_3$" characterizes the sector above $P_1$ and above $P_3$ and on both sides of $P_2$. ($a_i$ is the activation value of $v_i$, if $v_i$ represents the hyperplane $P_i$.)

Now layer $U_2$ is to define the sections. This means that each unit $v$ of layer $U_2$ represents one section, and the activation informs us whether the point is inside ($a_v = 1$) this section or not ($a_v = 0$). For each point only one unit of layer $U_2$ can be 1, the others are 0.

For describing that $a_u$ has to be taken positive, we define $W(u, v) = 1$ and for "NOT $a_u$" we use $W(u, v) = -1$. If a piece of information given by the unit $u$ does not matter, then we choose $W(u, v) = 0$. As already mentioned, the activation values are from the set $\{0, 1\}$. We define $b_v$ to be the number of units $u \in U_i$ that must have a positive activation for the sector that is described by $v \in U_{i+1}$. If e.g. $v$ describes a sector that is "above" the hyperplanes $P_1$, $P_3$ and $P_4$, but "below" $P_2$ and $P_5$, then we have $b_v = 3$. We choose $0.5 - b_v$ for the bias $\Theta_v$ of the unit $v$. Then we have an AND, because

$$a_v = \sum_{u \in U_i} W(u, v) \cdot a_u - b_v + 0.5 > 0 \qquad (9.1)$$

is fulfilled iff $W(u, v) \cdot a_u = 1$ for all $u \in U_i$ that represent a hyperplane where the sector is above the hyperplane.

### 9.2.2 The logical OR

Here we assume, that in the layer $U_2$, there is exactly one unit, that has the activation value 1 while the others are 0. This is true, because a point can only be situated in one section of the space, and this section is represented by the unit with the value 1.

Again we have $W(v, w) = 1$ for $a_v$ and $W(v, w) = 0$ if the unit $v$ does not matter. We do not have to consider a "NOT $a_v$", because we just want to combine the sectors that belong to the same class. With activation values being in $\{0; 1\}$, the network input function for a unit $w \in U_3 = U_O$ has the bias $-0.5$ for the OR. Then

$$a_w = \sum_{v \in U_2} W(v, w) \cdot a_v - 0.5 > 0 \qquad (9.2)$$

is fulfilled if $a_v > 0$ for a $v \in U_2$ with $W(v, w) = 1$.

Formula (9.2) can be translated. E.g. "$1 \cdot W(v_1, w) + 0 \cdot W(v_2, w) + 1 \cdot W(v_3, w) - 0.5 > 0$" simply means "If the point is in sector 1 or in sector 3, then it belongs to the class that is represented by $w$".

Now we have constructed the four layers as in figure 9.2, and with it we obtain a neural network that describes the correct classification of the space.

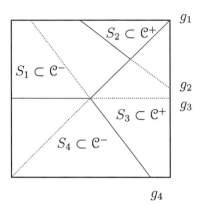

Figure 9.1:  An example for the piecewise division in the two-dimensional case

**Example 11**
*In figure 9.2, the net for the example of figure 9.1 is shown.*
*Line $g_3$ e.g. is parallel to the x-coordinate and therefore independent from x. This results in $w(u_{11}, u_{23}) = 0$.*
*Section $S_1$ is described as "above $g_1$, below $g_2$ and above $g_3$", but it can be found on both sides of $g_4$. Therefore we have $w(u_{24}, u_{31}) = 0$.*
*Figure 9.2 visualizes where the weights of W are 0 according to the dependencies of the example.*

### 9.2.3   Using $-1$ and $1$ instead of $0$ and $1$

If we use $[-1; 1]$ instead of $[0; 1]$ for the activations, then we just have to change the formulae for the calculation of the AND and the OR.
The values $x \in [0; 1]$ can be transformed into $x' \in [-1, 1]$ by

$$x' = \frac{x}{2} + \frac{1}{2}.$$

The following calculation transforms the equation for the separating hyperplane and leads us to the new weights and threshold values:

$$\sum_{i=1}^{n} a_{u_i} \cdot w(u_i, v) - \Theta = 0$$
$$\Leftrightarrow$$
$$2 \cdot \sum_{i=1}^{n} (a_i' - \tfrac{1}{2}) \cdot w(u_i, v) - \Theta = 0$$
$$\Leftrightarrow \sum_{i=1}^{n} a_i' \cdot w(u_i, v) - \tfrac{1}{2} \cdot (\sum_{i=1}^{n} w(u_i, v) + \Theta) = 0$$

Therefore when transforming the activation interval, we leave the weights $w(u_i, v)$ the same and transform the threshold from $\Theta$ into $\frac{1}{2} \cdot (\sum_{i=1}^{n} w(u_i, v) + \Theta)$.

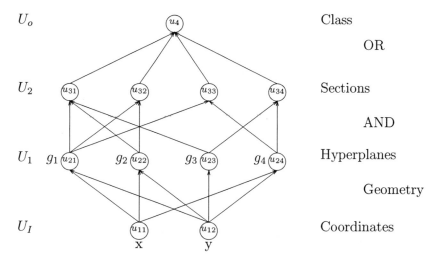

Figure 9.2: The network for the example of figure 9.1.

As the weights stay the same, for describing that $a_u$ has to be taken positive for the AND, we still define $W(u, v) = 1$ and for "NOT $a_u$" we have $W(u, v) = -1$, and if a piece of information given by the unit $u$ does not matter, then we choose $W(u, v) = 0$.

Let $m$ denote the number of units of the precedent layer that have to be taken into account, i.e. the number of weights between the two layers that equal $-1$ or $1$ but not $0$. When the network input function for a unit $u \in U_1$ has the bias $-m + 1$, then we the changeover between the two layers represents the logical AND, because

$$a_v = \sum_{u \in U_i} W(u, v) \cdot a_u - m + 1 > 0 \tag{9.3}$$

is fulfilled iff $W(u, v) \cdot a_u = 1$ for all $u \in U_1$.

To perform the logical OR, we need $W(v, w) = 1$ for $a_v$ and $W(v, w) = 0$ if the unit $v$ does not matter. Also here we do not have to consider a "NOT $a_v$". Let $c_w$ denote the number of sectors that belong to the class that is represented by the output unit $w$. In the case of the activation values belonging to $\{-1, 1\}$, the network input function for a unit $v \in U_2$ has the bias $c_w - 1$ for the OR. Then

$$a_w = \sum_{v \in U_2} W(v, w) \cdot a_v + c_w - 1 > 0 \tag{9.4}$$

is fulfilled if $a_v = 1$ for one $v \in U_2$ with $W(v, w) = 1$.

## 9.3   Disjunctive Normal Form

In the precedent sections we have seen, that units with their weights and threshold values are capable of representing boolean expressions. This can be used on one hand to structure the multilayer perceptron and on the other hand to extract it's structure.

Let us assume that an MLP is given that represents a boolean expression in the way described above. It is known, that every boolean expression can be transformed into a *disjunctive normal form* ([64]).

**Definition 11**

*1. Let $X_1, \ldots, X_n$ variables. We call a boolean expression $\alpha$ a disjunctive normal form (DNF), if it is a disjunction ($\bigvee$) of pairwise different conjunctions $X'_1 \wedge \cdots \wedge X'_n$ with $X'_i := X_i$, $X'_i := NOT\ X_i$ or $X'_i$ missing.*


*2. Two conjunctions $X = X'_1 \wedge \cdots \wedge X'_n\ Y = Y'_1 \wedge \cdots \wedge Y'_n$ are equal, iff $X'_i = Y_i$ for all $i \in \{1, n\}$. Otherwise they are different.*

It is obvious that the layers 2 to 4 of the multilayer perceptron, that we have constructed in section 9.2, uses a disjunctive normal form. We have the following four layers:

1. the input layer

2. the layer for the hyperplanes

3. the layer for the sectors (calculated out of layer 2 by the AND)

4. the output layer (calculated out of layer 3 by the OR)

Therefore during the construction of the multilayer perceptron, a boolean expression is transformed into an MLP. The other direction is still left to be considered:

We assume that we are given a multilayer perceptron. It has a threshold value function as activation function and the activations of the neurons of layer $U_1$ to $U_O$ are from $\{0; 1\}$. Then the way from layer $U_1$ to $U_O$ represents a boolean expression, and this boolean expression can be transformed into a disjunctive normal form:

$$\alpha = \bigvee_{\text{pairwise different}} (X'_1 \wedge \cdots \wedge X'_n).$$

We take the boolean expression of the given multilayer perceptron and transform it into a disjunctive normal form. Then we can represent this disjunctive normal form by another multilayer perceptron with four layers.

This method can be used to reduce the number of layers in a multilayer perceptron.

## 9.4 Conjunctive Normal Form

Similar considerations can be made about the conjunctive normal form.

**Definition 12**
1. *Let $X_1, \ldots, X_n$ variables. We call a boolean expression $\alpha$ a conjunctive normal form (CNF), if it is a conjunction ($\bigwedge$) of pairwise different disjunctions $X_1' \vee \cdots \vee X_n'$ with $X_i' := X_i$, $X_i' := NOT\ X_i$ or $X_i'$ missing.*

2. *Two disjunctions $X = X_1' \wedge \cdots \wedge X_n'\ Y = Y_1' \vee \cdots \vee Y_n'$ are equal, iff $X_i' = Y_i$ for all $i \in \{1, n\}$. Otherwise they are different.*

The same procedure of construction of a multilayer perceptron out of a disjunctive normal form can be done with the conjunctive normal form

$$\alpha = \bigwedge_{\text{pairwise different}} (X_1' \vee \cdots \vee X_n'),$$

but then the interpretation with the sectors does not hold any more. In this case the multilayer perceptron has four or five layers. The calculation of the OR has to be split up into two layers, because we can not assume that there is only one unit in $U_1$ that has the activation 1.

First we consider the case of the activations $\in \{-1; 1\}$. Here layer $U_2$ represents the OR, if we calculate its activations $a_v$ ($v \in U_2$) by

$$a_v := \begin{cases} 0 & \text{if } \sum_{u_i \in U_1} W(u_i, v) \cdot a_{u_i} + \frac{3}{2} - m \leq 0, \\ 1 & \text{if } \sum_{u_i \in U_1} W(u_i, v) \cdot a_{u_i} + \frac{3}{2} - m > 0, \end{cases}$$

with $m := |U_1|$ and $W(u_i, v) \in \{-1; 0; 1\}$ as in section 9.2.2. The AND can be calculated as described in section 9.2.1.

If the activations are in $\{0; 1\}$, then we need an additional layer. For each $a_u$ and each NOT $a_u$, that appear in the conjunctive normal form, we put one unit in layer $U_2$, so that we can have up to $2m$ units.

If the activation $a_v$ of a unit $v \in U_2$ represents $a_u$, then we choose $a_v := a_u$, and for NOT $a_u$ we choose

$$a_v := \begin{cases} 0 & \text{if } \sum_{u_i \in U_1} W(u_i, v) \cdot a_{u_i} + \frac{1}{2} \leq 0, \\ 1 & \text{if } \sum_{u_i \in U_1} W(u_i, v) \cdot a_{u_i} + \frac{1}{2} > 0, \end{cases}$$

with $W(u_i, v) := 0$ if $u_i \neq u$ and $W(u, v) := -1$. Now the activations for layer $U_3$ can be calculated by

$$a_w := \begin{cases} 0 & \text{if } \sum_{v_i \in U_2} W(v_i, w) \cdot a_{v_i} - \frac{1}{2} \leq 0, \\ 1 & \text{if } \sum_{v_i \in U_2} W(v_i, w) \cdot a_{v_i} - \frac{1}{2} > 0, \end{cases}$$

with $W(v_i, w) := 1$ for all $v_i$. This gives the OR, and the AND can be constructed as in section 9.2.1.

## 9.5   Using a Sigmoidal Activation Function

In the classical definition of a multilayer perceptron, there is a sigmoidal activation function. This means, that the activation of a unit is calculated from the activations of the neurons of the previous layer not with a threshold activation function $f : \mathbb{R} \to \{0; 1\}$, resp. $f : \mathbb{R} \to \{-1; 1\}$, but there is an activation function $f : \mathbb{R} \to [0; 1]$ resp. $f : \mathbb{R} \to [-1; 1]$ applied:

$$a_v := f(\sum_{i=1}^{n} W(u_i, v) \cdot a_{u_i} - \Theta),$$

with $f$ being a sigmoidal function:

- $f(x) = \frac{1}{1+e^{-\beta x}}$, $\beta > 0$ (logistic function, asymptotic at 0 for $x \to -\infty$ and at 1 for $x \to \infty$)

- $f(x) = tanh(\beta x) = \frac{e^{\beta x} - e^{-\beta x}}{e^{\beta x} + e^{-\beta x}}$ (asymptotic at $-1$ and 1)

- $f(x) = \frac{1}{\pi}(\frac{\pi}{2} + arctan(\beta x))$ (asymptotic at 0 and 1)

In these three functions, the parameter $\beta$ specifies how steep the function is.

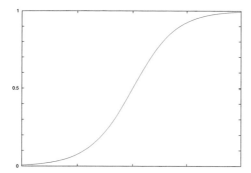

In this chapter we took a special case by assuming that we have an activation function that is that steep that it is a threshold value function. This way we receive for each unit an activation that is 0 or 1 (resp. $-1$ or 1), that can be interpreted as a sharp distinction between two opposite statements.

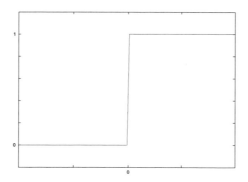

If we have a sigmoidal activation function, then the activation of the unit $v$ also gives a sort of membership degree for the input point to the side it belongs to: The bigger the distance $|a_v - 0.5|$ is, the more is the point away from the separation plane, and the more it belongs either to the side $S_v^+$ or to the side $S_v^-$.

## 9.6 Deriving a Multilayer Perceptron from Cluster Prototypes

In chapter 7, we already described how to interpret the prototypes of fuzzy clusters as the basis for defining separating hyperplanes between the clusters. The approach to form the separating hyperplanes was to place the hyperplanes in the middle between two prototypes. For $c$ clusters, this procedure normally results into $\binom{c}{2}$ hyperplanes, when we use each possible pair of prototypes to define a hyperplane, but we can reduce this number.

Using all these hyperplanes would result in a large number of hyperplanes and therefore in a large number of units in the first hidden layer of the resulting multilayer perceptron. For this reason we need to reduce the number of hyperplanes as far as possible and take into account only the hyperplanes that are necessary for the classification, the *relevant* hyperplanes.

**Definition 13**

Let $c_1, \ldots . c_c$ be the set of prototypes in the data space $D = [a_1, b_1] \times \cdots \times [a_n, b_n]$. Let $A$ and $B$ be two points in the data space, either data or prototypes. Let $d(A, B)$ be the distance between $A$ and $B$.

1. The hyperplane $H_{ij}$ between the clusters $c_i$ and $c_j$ is called relevant for $c_i$ if there is a point $P \in H_{ij} \cap D$ so that

$$d(P, c_i) = \min\{d(c_k, P) \mid k = 1, \ldots, c\}.$$

2. The hyperplane $H_{ij}$ between the clusters $c_i$ and $c_j$ is called irrelevant for $c_i$ if it is not relevant for $c_i$.

**Corollary 6**

If the hyperplane $H_{ij}$ is relevant for $c_i$, then it is also relevant for $c_j$. We call the hyperplane $H_{ij}$ relevant.

**Proof:** If $H_{ij}$ is relevant for the cluster $c_i$, then there is a point $P \in H_{ij} \cap D$ with $d(P, c_i) = \min\{d(c_k, P) \mid k = 1, \ldots, c\}$. As $P \in H_{ij}$, we have $d(P, c_i) = d(P, c_j)$ and therefore $d(P, c_j) = \min\{d(c_k, P) \mid k = 1, \ldots, c\}$. Then $H_{ij}$ is relevant for $c_j$.

If $H_{ij}$ is irrelevant in $c_i$, then for all points $P \in H_{ij}$ there is a cluster $c_k$ with $d(P, c_k) < d(c_j, P)$. From $d(P, c_k) < d(c_i, P) = c(c_j, P)$ follows that $P$ is also irrelevant for $c_j$.                                                          □

**Example 12**

*Figure 9.3 shows an example with one irrelevant hyperplane. The drawing illustrates that hyperplane $H_{13}$ is irrelevant, because the points that belong to $H_{13}$ are separated from cluster $c_1$ by $H_{12}$ and and from cluster $c_3$ by $H_{23}$. They have a greater membership degree to $c_2$ than to $c_1$ as stated by $H_{12}$, and a greater membership to $c_2$ than to $c_3$ as stated by $H_{23}$. As the points of $H_{12}$ belong neither to $c_1$ nor to $c_3$, the hyperplane is irrelevant.*

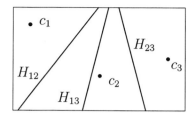

Figure 9.3: An example for an irrelevant hyperplane

## Remark 22

*When constructing an MLP we do not have to consider irrelevant hyperplanes for the classification.*

*A hyperplane $H_{ij}$ is irrelevant, iff for all $P \in H_{ij}$, there is a cluster $c_k$, $k \neq i, j$, so that*

$$d(P, c_k) < d(P, c_i) = d(P, c_j).$$

*This means, that the point $P$ belongs to $c_k$ (or another even closer cluster), but not to $c_i$ or $c_j$ anymore. The classification decision is taken by other hyperplanes, and therefore we do not need to consider irrelevant hyperplanes for our construction.*

When transforming a clustering result into an MLP, there are often several hyperplanes that are not relevant, because they are in a similar situation as $H_{13}$ in the above example. Now we have to figure out, which are these irrelevant hyperplanes. The principle behind the procedure is the following: To each cluster $c_i$ belongs a section, that is separated by a number of planes from the other clusters. These hyperplanes form a subset of $\{H_{i1}, \ldots, H_{i(i-1)}, H_{i(i+1)}, \ldots, H_{ic}\}$. These hyperplanes are those that separate cluster $c_i$ from the other clusters. Now we have to figure out which hyperplanes are irrelevant.

## Definition 14

Let $\mathcal{H}_i := \{H_{jk} | j = i \text{ or } k = i\}$ and $D$ be the data space. Let

$$S_i := \{P \in D | d(P, c_i) < d(P, c_j), j \neq i\}.$$

*Then $S_i$ is a section of the dataspace. $S_i$ is called the section belonging to $i$.*

If a hyperplane $H_{ij}$ is relevant, than it is relevant in at least one corner of the polyhedron that defines the section belonging to $c_i$ or the section belonging to $c_j$.

We can calculate all the potential corners by forming all possible intersection points of the hyperplanes of $H$ and all the hyperplanes formed by the outer boundaries of the data space. For each of these corners we have to check the membership degrees to the clusters. If it has the biggest membership degree for the cluster to which the section belongs, than the corner is relevant, and therefore all those hyperplanes that intersect in this corner are relevant. If there is another cluster $c_j$ with $\mu_j(P) > \mu_i(P)$, than the point $P$ and a neighborhood of $P$ belong to another cluster, and therefore the point is irrelevant.

**Definition 15**
Let $P \in H_{ij}$ with $H_{ij}$ being a separating hyperplane in the n-dimensional data space. $P$ is called an intersection point of $H_{ij}$ iff there are $n-1$ planes $\bar{H}_1, \ldots, \bar{H}_{n-1}$ with the $\bar{H}_k$ being either one of the separating hyperplanes or one of the data space boundaries, so that $P = H_{ij} \cap \bar{H}_1 \cap \ldots \cap \bar{H}_{n-1}$.

We only have to check the separation inside the dataspace, not outside. Thus we only have to examine the intersection points inside the data space.

**Remark 23**
*It is important to define the data space as small as possible to exclude cases where hyperplanes are relevant only in an area where we do not have any data to be classified.*

A hyperplane is relevant if only one of it's points is relevant, and it is irrelevant if none of its intersection points turns out to be relevant. But checking all intersections points where a hyperplane belongs to until we find a relevant intersection point or until we reach the end, takes too long.
There is a possible simplification. When having the intersection $I$ of $k < n$ hyperplanes in the $n$-dimensional space, we have to consider these k hyperplanes together with $n - k$ boundaries of the data space. But we do not have to check all possible choices of boundaries, but only one is sufficient. There are several possibilities for the resulting point when we intersect these $n$ hyperplanes with each other:

1. The resulting intersection point $P$ is relevant. Then the hyperplane is relevant.

2. The resulting intersection point $P$ is irrelevant.

   (a) The whole intersection is irrelevant, i.e. there is no intersection point of the intersection of the $k$ hyperplanes with any of the $n-k$ boundaries.

(b) There is a point $Q$ of $I$ that is relevant. This means, that there must be another separating hyperplane $h$ not being involved in $I$, that is covering $P$, because it is situated between $P$ and $c_i$.

As it has to be behind $Q$ although it is in front of $P$ (relatively to $c_i$), it has to cross $I$ somewhere. This crossing point then is another intersection $I \cap h$ that is to be examined at a later step in the algorithm.

Therefore we only have to consider an arbitrary choice of boundaries, not all possible combinations of boundaries.

There are two possible directions for the procedure of choosing $n$ separating hyperplanes and data space boundaries: we can either start with $n$ separating hyperplanes and then reduce the number of separating hyperplanes and increase the number of boundaries, or we can start with single hyperplanes and $n-1$ boundaries and then increase the number of separating hyperplanes and decrease the number of boundaries.

The reason for starting with $n$ and then reducing is, that during the procedure we can note, which hyperplanes we have already found out to be relevant. Then we only have to check those intersections where at least one hyperplane is not already detected to be relevant. And with the intersections including as many hyperplanes as possible, we can more quickly fill up our list of relevant hyperplanes.

**Remark 24**

*We do not have to consider all possible combinations of hyperplanes. Instead we can choose one cluster $c_i$ after the other and examine only the hyperplanes $H_{ij}$ and $H_{ki}$ that belong to $c_i$ and their combinations with each other and with the data space boundaries. Then we can examine whether the hyperplane is relevant w.r.t. definition 13.*

Now we have chosen the relevant separating hyperplanes and constructed a multilayer perceptron with as few neurons in the first hidden layer as possible.

The multilayer perceptron can be constructed by knowing the relevant hyperplanes and the two clusters they are formed by. The weights first hidden layer are assigned by using the equation for the hyperplanes. The second hidden layer is formed by the AND as described in section 9.2.1 while the last layer is formed by the OR as to be found in section 9.2.2.

The resulting MLP can be trained to improve and reduce the error.

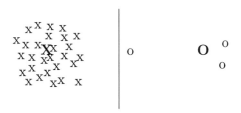

Figure 9.4: An example for the advantage of a constructed MLP

**Remark 25**

*Note that it may occur that the error of a trained multilayer perceptron is less than that one of a constructed MLP, but the classification of the latter performs better. Figure 9.4 shows an example for such a case. The separating line was constructed in the middle between the two clusters.*

*If the MLP would be trained, the separating line would move to the right to decrease the error. The error for the data of the left cluster would be less, when the separation moved, and there are only few data in the right cluster that could lower the overall error.*

Until here we only considered an MLP that reproduces the solution of a piecewise linear separable classification problem, but we are also able to use our technique to adapt the system to a continuous output. We will describe this in section 10.1.4.

**Remark 26**

*The problem of finding the relevant planes between the different prototypes is the same as finding a Voronoi diagram of the prototypes or a Delaunay triangulation of the separating hyperplanes ([25, 36]).   E.g. up to eight dimensions the Quickhull algorithm can be used ([10]).*

**Remark 27**

*The overall construction of the multilayer perceptron out of the prototypes is deterministic.*

In the following we want to demonstrate the methods of this chapter on an example.

## 9.7 Constructing a Multilayer Perceptron for the Iris Data Set

In section 7.3 we considered the classification of the iris data set. We used only three dimensional of the data space. The approach to form the separating hyperplanes was to place the planes in the middle between two prototypes. For $c$ clusters, this procedure normally results into $\binom{c}{2}$ hyperplanes, when we us each pair of prototypes to define a hyperplane, but we can reduce this number.

### 9.7.1 Determining the hyperplanes

The iris data as described in section 7.3 set was clustered with five clusters. We received the following prototypes.

$$c_1 = \begin{pmatrix} 3.43 \\ 1.46 \\ 0.25 \end{pmatrix}, c_2 = \begin{pmatrix} 2.52 \\ 3.82 \\ 1.15 \end{pmatrix}, c_3 = \begin{pmatrix} 2.93 \\ 4.53 \\ 1.43 \end{pmatrix}, c_4 = \begin{pmatrix} 2.89 \\ 5.20 \\ 1.93 \end{pmatrix}, c_5 = \begin{pmatrix} 3.13 \\ 5.99 \\ 2.16 \end{pmatrix}.$$

To train a multilayer perceptron with the data, we apply a linear transformation on the data space to receive input values with mean value 0 and standard deviation 1. Then the prototypes are the following:

$$c_1 = \begin{pmatrix} 0.65 \\ -1.31 \\ -1.25 \end{pmatrix}, c_2 = \begin{pmatrix} -1.24 \\ 0.04 \\ -0.06 \end{pmatrix}, c_3 = \begin{pmatrix} -0.29 \\ 0.44 \\ 0.3 \end{pmatrix}, c_4 = \begin{pmatrix} -0.39 \\ 0.82 \\ 0.96 \end{pmatrix}, c_5 = \begin{pmatrix} 0.17 \\ 1.27 \\ 1.26 \end{pmatrix}.$$

In the following we will visualize the data with this transformed data set.

When we needed all the hyperplanes defined by two clusters at a time, we would end up with $\binom{5}{2} = 10$ hyperplanes, quite a large number. We have a look at the visualization to see, which hyperplanes are superfluous.

First of all we consider the first cluster, that consists of all the points of class $\mathcal{C}$. In figure 9.5 we see the two hyperplanes that are needed, i.e. $H_{12}$ and $H_{13}$.

In figure 9.6, the red hyperplanes $H_{14}$ and $H_{15}$ are superfluous. As these hyperplanes are situated behind $H_{12}$ and $H_{13}$, those points, that can be classified by them, are already classified as 'not belonging to $c_1$ by $H_{12}$ and $H_{13}$.

Therefore we have $A_1 = \{c_2, c_3\}$. We continue with $c_2$ and find out that we can neglect $H_{25}$, because it is situated behind $H_{24}$ apart from the small area there no input points are situated, so that we can neglect this. Finally, we have $A_2 = \{c_1, c_3, c_4\}$.

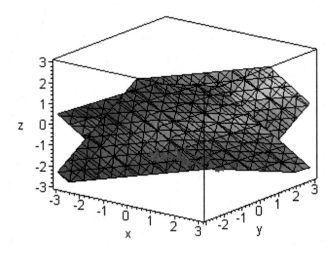

Figure 9.5: The separating planes needed for $\mathcal{C}_1$ .

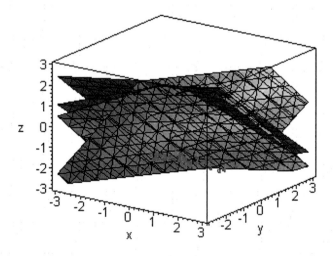

Figure 9.6: The red separating planes are superfluous for $\mathcal{C}_1$

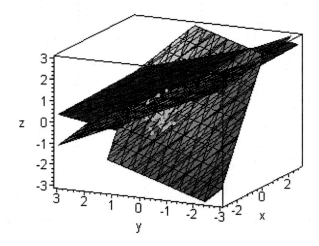

Figure 9.7: The red separating plane is superfluous for the second cluster.

We continue and receive $A_3 = \{c_2, c_4\}$, $A_4 = \{c_2, c_3, c_5\}$ and $A_5 = \{c_3, c_4\}$. In the end, we know that we need the hyperplanes $H_{12}$, $H_{13}$, $H_{23}$, $H_{24}$, $H_{34}$, $H_{35}$, $H_{45}$.

**Remark 28**

*The heuristical approach could show us, that the clusters can already be classified when using only $H_{12}, H_{23}, H_{34}, H_{45}$. The reason is that the input data are not distributed on the whole data space, so that the areas where no input data are situated do not have to be classified correctly in respect of the clusters.*

Now we build up the multilayer perceptron with these informations. The first layer is to represent the hyperplanes. We use seven units, one for each hyperplane. The following layer is to present the logical AND, and the output layer is formed by the logical OR. In figure 9.8, you can see how the weights have to be chosen.

We perform the tool with mmlpt (written by C.Borgelt, Software mmlpt and mmlpx, http://fuzzy.cs.uni-magdeburg.de/borgelt/software.html) choosing a steep sigmoidal function ($f(x) = \frac{1}{e^{-10 \cdot x}}$) to receive a clear classification.

We have six misclassified data. Figure 9.9 shows the misclassified data of the first class $\mathcal{C}_1$, figure 9.9 and figure 9.9 those of the other classes. These

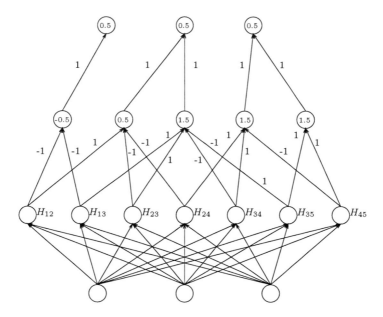

Figure 9.8: The constructed MLP for the iris data set

misclassifications result simply from the fact that these points are closer to the wrong prototype than to the correct one, so that they are sorted for the wrong cluster. Therefore the six misclassified data are exactly what we expected.

## 9.7.2   Simplifying the Multilayer Perceptron

In section 9.3 we already mentioned, that an MLP can be simplified by the means of the boolean expressions. We want to show this with an example. We use the same data as in the precedent section, but we have an arbitrary MLP that learns the values.

Let a multilayer perceptron with three hidden layers consisting of five units be trained for the iris data set. We train the MLP with mmlpt. After 1000 iterations, the resulting MLP has a root mean squared error of 0.16181.

In the first hidden layer each unit still represents one of five separating hyperplanes in the data space, while the change from the first hidden layer to the output layer represent a boolean expression, but there is no structure like a disjunctive or conjunctive normal form. We want to transform it into such a form.

For this purpose, we remove the first hidden layer. It is not involved in the

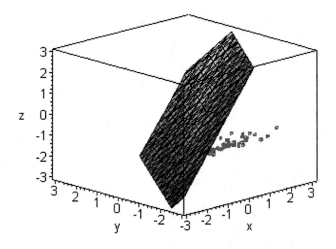

Figure 9.9: Class $\mathcal{C}_1$ has one misclassified data.

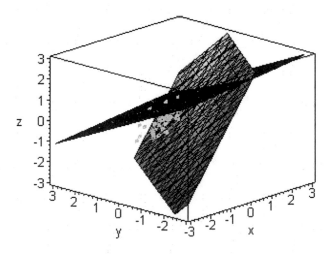

Figure 9.10: Class $\mathcal{C}_2$ has three misclassified date.

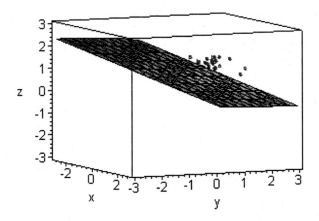

Figure 9.11: Class $\mathcal{C}_1$ has two misclassified date.

calculation of the boolean expression, so that we do not consider it for the simplification. The resulting net uses the first hidden layer of the original net as input layer. We still use activations of the units that are in $\{-1; 1\}$ and a sigmoidal activation function.

As the input data are representing boolean inputs, we construct a data set with five boolean inputs and classify these data with the reduced MLP. Table 9.1 shows the results in the middle. On the right, we have transformed the output data $y_i$ into boolean values $y_i'$ by

$$y_i' := \begin{cases} 0 & \text{if } y_i < 0.5 \\ 1 & \text{if } y_i \geq 0.5 \end{cases}$$

We examine each output unit separately. By using the QuineMcCluskey-Algorithm [85, 66], we can reduce the three boolean expression $Y_1, Y_2, Y_3$ that are represented by the output units of the MLP to the following disjunctive normal forms:

| Boolean input values | | | | | Output values | | | Transformed boolean output | | |
|---|---|---|---|---|---|---|---|---|---|---|
| 0 | 0 | 0 | 0 | 0 | 0.990098 | 0.0159504 | 0.000269146 | 1 | 0 | 0 |
| 0 | 0 | 0 | 0 | 1 | 0.990067 | 0.016001 | 0.00026908 | 1 | 0 | 0 |
| 0 | 0 | 0 | 1 | 0 | 3.72242e-05 | 0.994703 | 0.00505888 | 0 | 1 | 0 |
| 0 | 0 | 0 | 1 | 1 | 0.421091 | 7.64677e-07 | 0.861719 | 0 | 0 | 1 |
| 0 | 0 | 1 | 0 | 0 | 0.000857109 | 0.999712 | 9.10872e-05 | 0 | 1 | 0 |
| 0 | 0 | 1 | 0 | 1 | 0.00073579 | 0.999754 | 8.98866e-05 | 0 | 1 | 0 |
| 0 | 0 | 1 | 1 | 0 | 0.000365814 | 1.00974e-05 | 0.992443 | 0 | 0 | 1 |
| 0 | 0 | 1 | 1 | 1 | 0.0574019 | 4.67001e-08 | 0.995296 | 0 | 0 | 1 |
| 0 | 1 | 0 | 0 | 0 | 0.0414961 | 0.982561 | 0.000130566 | 0 | 1 | 0 |
| 0 | 1 | 0 | 0 | 1 | 0.997837 | 0.00323777 | 0.000309967 | 1 | 0 | 0 |
| 0 | 1 | 0 | 1 | 0 | 0.000624629 | 0.999795 | 8.79493e-05 | 0 | 1 | 0 |
| 0 | 1 | 0 | 1 | 1 | 0.000630439 | 0.999793 | 8.80269e-05 | 0 | 1 | 0 |
| 0 | 1 | 1 | 0 | 0 | 0.996318 | 0.00566176 | 0.000294936 | 1 | 0 | 0 |
| 0 | 1 | 1 | 0 | 1 | 0.968752 | 0.0526555 | 0.000241343 | 1 | 0 | 0 |
| 0 | 1 | 1 | 1 | 0 | 0.0108923 | 3.82021e-06 | 0.96717 | 0 | 0 | 1 |
| 0 | 1 | 1 | 1 | 1 | 0.0616362 | 4.81245e-08 | 0.994954 | 0 | 0 | 1 |
| 1 | 0 | 0 | 0 | 0 | 0.974783 | 0.0421871 | 0.000246433 | 1 | 0 | 0 |
| 1 | 0 | 0 | 0 | 1 | 0.990069 | 0.0159983 | 0.000269084 | 1 | 0 | 0 |
| 1 | 0 | 0 | 1 | 0 | 6.55168e-08 | 0.0874694 | 0.982357 | 0 | 0 | 1 |
| 1 | 0 | 0 | 1 | 1 | 0.030288 | 9.36811e-08 | 0.994994 | 0 | 0 | 1 |
| 1 | 0 | 1 | 0 | 0 | 0.360572 | 0.790267 | 0.000166234 | 0 | 1 | 0 |
| 1 | 0 | 1 | 0 | 1 | 0.000952159 | 0.999677 | 9.21574e-05 | 0 | 1 | 0 |
| 1 | 0 | 1 | 1 | 0 | 6.99737e-08 | 0.0764568 | 0.983363 | 0 | 0 | 1 |
| 1 | 0 | 1 | 1 | 1 | 0.058504 | 4.56895e-08 | 0.995299 | 0 | 0 | 1 |
| 1 | 1 | 0 | 0 | 0 | 0.00061147 | 0.999791 | 9.04647e-05 | 0 | 1 | 0 |
| 1 | 1 | 0 | 0 | 1 | 0.997445 | 0.00384605 | 0.000305789 | 1 | 0 | 0 |
| 1 | 1 | 0 | 1 | 0 | 0.000629433 | 0.999793 | 8.80184e-05 | 0 | 1 | 0 |
| 1 | 1 | 0 | 1 | 1 | 0.000161838 | 0.999009 | 0.000620712 | 0 | 1 | 0 |
| 1 | 1 | 1 | 0 | 0 | 0.996808 | 0.0048575 | 0.000299611 | 1 | 0 | 0 |
| 1 | 1 | 1 | 0 | 1 | 0.997459 | 0.00382346 | 0.000305999 | 1 | 0 | 0 |
| 1 | 1 | 1 | 1 | 0 | 0.00409777 | 7.90619e-07 | 0.993961 | 0 | 0 | 1 |
| 1 | 1 | 1 | 1 | 1 | 0.0505989 | 5.33603e-08 | 0.995237 | 0 | 0 | 1 |

Table 9.1: The Input and Output Data of the MLP without the first layer.

$$Y_1 = \bar{x}_2\bar{x}_3\bar{x}_4 \vee x_2x_3\bar{x}_4 \vee x_2\bar{x}_3\bar{x}_4x_5$$
$$Y_2 = \bar{x}_2x_3\bar{x}_4 \vee x_2\bar{x}_3x_4 \vee x_2\bar{x}_3\bar{x}_4\bar{x}_5 \qquad (9.5)$$
$$Y_3 = \bar{x}_1\bar{x}_2\bar{x}_3x_4x_5 \vee x_1\bar{x}_2\bar{x}_3x_4 \vee x_3x_4$$

Now we can use the disjunctive normal forms (one for each output unit) to construct a smaller multilayer perceptron. The first layer is the first layer of the trained MLP represents the separating hyperplanes. we did not change this at all.

The second layer results from the logical AND and has nine units, each one representing one of the conjunctive expressions in (9.5). The third layer is the output layer and represents the logical OR, the disjunction.

Figure 9.12 shows the resulting MLP. It has less units and one layer less, and many of its weights equal 0. E.g. is the first hyperplane only needed for the calculation of two conjunctive expression. Therefore classifying data with this constructed MLP is faster that with the originally trained one, while the results differ only negligibly.

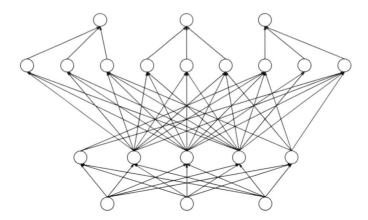

Figure 9.12: The reduced MLP for the Iris data set.

## 9.8   Further considerations

We have seen a possible interpretation of a multilayer perceptron, with a threshold value function as activation function. This interpretation divides the space into sections by using hyperplanes. This means that every piecewise linearly separable problem can be described by such a multilayer perceptron.

Because of having a precise construction, we can transform a geometrical description (by hyperplanes) into a multilayer perceptron. And this is the fact where the fuzzy classification system with the Łukasiewicz-t-norm and the multilayer perceptron meet.

Both of them classify the data space by separating hyperplanes that result in different sectors. And these sectors can be assigned to the classes. These similarities can be used in two different ways:

The first case where this can be useful is the situation, when we have already a rough description of the classes e.g. from a fuzzy classification system or from fuzzy clusters. We can initialize the neural network with the known values and then improve the net by learning. This at least reduces the learning time that the multilayer perceptron needs and will also lead to better solutions in the classification, if the problem is more complicated and the neural network gets stuck in local minima easily.

Vice versa, we can have a look into the 'black box' of the neural network. As an arbitrary multilayer perceptron does not have the same formal appearance as the constructed MLP in the previous sections, the sections cannot be derived directly from the neural network. The first hidden layer can still be interpreted as a representation of the separating hyperplanes, but the following layers normally do not represent an AND or an OR- function. (Different approaches to neural networks using AND and OR can be found [4].)

It is not possible to adapt the backpropagation algorithm to the formalities of the boolean expressions, because then the weight would have to change by the large steps of 0.5 or 1. Thus we need to pass by the method used in section 9.7.2. We have to extract the boolean expression represented by the second hidden layer to the output layer, derive the disjunctive normal form of this expression e.g. by using the QuineMcCluskey-algorithm, and build a new MLP that is representing the same boolean expression. And the resulting multilayer perceptron can be interpreted as a representation of sections formed by separating hyperplanes.

When we have a multilayer perceptron that succeeds in classifying our data, then we can transform the neural network into a fuzzy classification system, passing by the visualization. As such a fuzzy classification system consists of rules with fuzzy sets that can be transformed into linguistic values, we finally receive a linguistic description of the classification. Therefore the multilayer perceptron turns out to be interpretable.

# Chapter 10

# An Application: Influence of Weather Data on Aircraft delays

In this chapter we want to demonstrate an application of the method that we have derived in the previous chapters. This will not only show how to use the algorithms but also demonstrate some of its strengths.

Parallel to the rapidly increasing global economy interweavement the demand for air transportation capacity increases. In order to minimize transport delays in the air traffic network a major effort was made since ever to mitigate weather effects on air transport as one important contributor to air traffic system disturbances.

It is obvious that bad weather conditions lead to increase in staggering of aircraft. But weather (being a multi-factor system itself) is not the only factor to be taken into account for predicting the overall amount of delay in a concrete air traffic situation. Available runway capacity, controller and pilot workload, quality of the technical infrastructure and mix of different aircraft types as well as airspace organization and the applicable operational rules are examples of other important factors that may influence actual delay in addition to weather effects.

Any improvement of knowledge of the factors influencing the delay helps the air transport stackholders to plan and manage the air traffic in a more efficient and predictable way. Air traffic control, airport authorities and the airlines will use those information to improve their local system with positive effects on the global network [78] to increase airport capacities and reduce delay times.

Capacity studies are performed by using simulation tools and by varying parameters that are known to influence the airport capacity. It is well known

| ATIS | AFW |
|---|---|
| air pressure(hpa) | air pressure(hpa) |
| temperature(deg $C$) | wind speed(KT) |
| visibility (m) | wind direction |
| cloud amount | |
| height of clouds(tft) | |
| precipitation | |
| intensity of precipitation | |
| wind speed (KT) | |
| wind direction | |

Table 10.1: The attributes of the data sets providing weather information

that the traffic influences the delay significantly [2, 27, 84]. Examining the weather data's influence on the delay leads to further understanding of the conditions.

## 10.1   The dataset

In [87], for a data set for inbound traffic at Frankfurt, the influence of weather conditions on delays was examined, using different statistical methods. Fuzzy clustering and multilayer perceptrons belonged to these methods. The aim of this work was to examine which attributes influence the delay and how strong this influence turns out to be.

The questions occurs whether an MLP can be constructed on the basis of the clustering results, that can improve the clustering results by further learning. The first flight activity at Frankfurt took place in 1785 in the form of balloon starts. At its current location, Frankfurt airport started working in 1936. The main location factors were the good transportation connections and the main wind direction [3]. Nowadays it is the airport with the largest passenger volume in Germany.

The examined dataset consists of measured data taken at Frankfurt airport in 1998 and 1999, i.e. the weather attributes and the corresponding starts and arrivals with the delay time as described in [87, 98]. Current weather data are provided by the "Deutsche Wetterdienst" (DWD) via the "Automatic Terminal Information Service" (ATIS) and with the "aerodome meteorological conditions" (Aktuelles Flugplatzwetter, AFW) taken at the airport. The attributes of these datasets that are used for the analysis of the weather data are listed in table 10.1.

| **Traffic Data** |
| --- |
| date and time, accuracy in seconds (Arrival/Departure/TMA) type of aircraft class of aircraft (light, medium, heavy) the runway for the arrival/departure |

Table 10.2: The attributes of the data sets providing aircraft information

The weather data of ATIS are normally taken every 30 minutes. Only in times of rapid changes they are taken more often. The AFW dataset consists of measures that are taken every minute.

Sometimes missing values occur in the ATIS data set, when e.g. measuring instruments fail. In 1998 there were more than 18.000 measurements including about 2000 incomplete ones. In the case of missing air pressure or wind data, the data set could be completed by using the AFW measures.

The information given about the arrivals and departures is shown in table 10.2. The Terminal Manoevring Area (TMA) is the region close to the airport where the aircrafts are served by the airports air traffic controllers. The TMA-time is the time when the aircraft is entering the TMA. In Frankfurt an aircraft normally needs from this point of time until the arrival approximately 30 minutes.

In [87] it was examined, how long it took the aircrafts to pass the TMA. Then the delays that are already caused elsewhere (e.g. at the departure airport of the aircraft) do not influence the results and the weather influences can be examined separately.

## 10.1.1 Preparing the Data

[87] performs a principle component analysis. The result can be visualized with the first three dimensions. This leads to the fact that the data separate very clearly into two sets: the data for clouded weather and those for cloudless sky.

Therefore in the following we will consider these two cases separately as different data sets.

The next thing to be done was the regression analysis. It is already well known that the number of aircrafts being on arrival at the airport has a great influence on these times. The regression analysis ([87, 92]) can be used to adopt the travel time as shown in figure 10.1.

In the left picture of figure 10.1 we can see how the travel time depends on the traffic. In the right picture the corrected travel time is drawn in red,

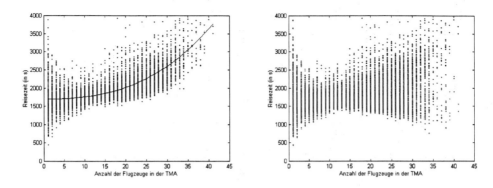

Figure 10.1: Influence of the traffic on the travel time

while for better comparison the blue points show the original data.

In [87] is also described how the relevant attributes are chosen, so that we consider 10 attributes in the case of clouded weather and 6 in the case of cloudless weather.

In the following we will examine the corrected travel time, because we are only interested in the influence of the weather. The influence of the traffic was already examined in other works ([98]).

## 10.1.2   The Clustering

The aim of clustering the weather data was to find structures that lead towards a prediction of delay. In [87], the fuzzy c-means algorithm was performed on the weather data resulting in eight clusters. The prototypes were only slightly different each time the clustering was done. Therefore the prototypes can be assumed to be as given as listed in the appendix.

Assigning the mean to each cluster leads to a prediction system. The resulting RMSE (Root mean squared error) is listed in table 10.3.

## 10.1.3   Construction of the MLP

Now we want to construct an MLP out of the prototypes of the clustering that performs nearly the same classification into clusters and the same outputs. As we have 8 clusters, there are $\binom{8}{2} = 28$ possible separating hyperplanes. We apply the reduction method described in section 9.6 to check whether all the hyperplanes are relevant or not.

In the case of cloudless sky, the number of necessary hyperplanes reduces by one, in the case of cloudy sky, all 28 hyperplanes are needed.

Then the multilayer pereptron can be constructed as described in chapter 9. The resulting MLP has 8 output units representing the clusters. When the input data is one of the prototypes, then the output unit representing its cluster adopt a value close to 1 while the others have an activation that is near 0.

As the clustering system is predicting a continual value instead of the resulting value, we have to add another layer as described in section 10.1.4.

**Remark 29**
*For the construction of the MLP it does not matter whether the clustering was performed on the whole dataspace containing input and output values of the problem, or whether it was performed only on the input data and the output data were determined by another method, as e.g. the mean as in this case.*

*The constructed MLP is a system that is duplicating the function of the clustering system.*

## 10.1.4  Determining the weights for a continuous output

The algorithm presented in chapter 9 results in an MLP that assigns the data to a cluster. If the input date $p_i$ is the prototype of cluster $c_i$, the output of the unit $u_i$ representing the cluster $c_i$ assumes a value close to 1 while the others are close to 0. As there has to be a single output unit to present the continuous output, we need one additional layer.

Therefore we assume that the constructed MLP has to assign the output $o_i$ to each prototype $p_i$. Calculating backwards, we determine the weights for the net input function of the output neuron.

The clustering output $o_i$ of each prototype $p_i$ was set to the desired output value of the cluster belonging to the prototype. We use the same principle. For each prototype $p_i$ we can calculate

$$s_i := f^{-1}(o_i); \; i = 1, \ldots, c,$$

with $f$ being the logistic function and $o_i$ being the desired output value for $p_i$.

$$f(x) = \frac{1}{1 + e^{-x}} \Leftrightarrow f^{-1}(x) = -ln(\frac{1}{x} - 1).$$

We solve the linear system of equations

$$s_i = \sum_{j=1}^{c} \alpha_j a_j^{(p_i)}, i = 1, \ldots, c,$$

with $a_j(p_i)$ being the activation of the unit representing the $c_j$ when the input date is $p_i$. The results $\alpha_i$, $i = 1, \ldots, c$, of the system are the weights for the single output unit.

The prediction value for the delay is determined by scaling the output interval $[0, 1]$ of the MLP to the range of the data set.

This construction of the continous output value offers new possibilities of interpolation. This is not used for classical classification problems, but it can be helpful to use clustering results as in our application shown.

It is possible to initialize an MLP with expert knowledge that gives just distinct values and therefore represents a classification. But still the learning process can turn the system into a approximation system by training the continous output.

The output of the constructed NLP is not identical with the output of the clustering system. The reasons for this can be found in the following section.

## 10.2   The Results

To judge the performance of the constructing method we compared the RMSE of the clustering to the RMSE of the constructed MLP and to the RMSE of the MLP when it was improved by learning.

In table 10.3 we list up the different RMSEs for the clustering and the MLP. The first line shows the RMSE that the clustering of [87] resulted in.

The second line lists up the RMSE that the MLP has directly after it has been constructed and without any further learning. We can see that it performs $1.2 - 8.4\%$ less than the clustering it was derived from.

The following section in the table shows how many learning epoches the MLP needs to perform as good as the clusters do.

The third part of the table show how the MLP does after it has been learning for 1000 epoches. Now it is between 1.2 and 7.9% better than the clustering. The MLP that is constructed out of the prototypes is slightly less well performing than the clustering itself. This can easily be explained. For this purpose, let us consider the membership functions of the clustering and the activation function of the MLP.

The membership function of the fuzzy clustering as in chapter 6.1 is defined by

$$\sum_{i=1}^{C} \sum_{j=1}^{N} u_{ij}^m d_{ij}$$

with $d_{ij} = ||v_i - x_j||^2$ being the Euclidian distance between the prototype $v_i$

| | clouded 1998 | clouded 1999 | cloudless 1998 | cloudless 1999 |
|---|---|---|---|---|
| **RMSE Clustering** | 231.72 | 317.63 | 167.61 | 214.27 |
| **RMSE Constructed MLP** **Relation to Clustering** | 239.99 -3.6% | 344.22 -8.4% | 171.13 -2.1% | 216.86 -1.2% |
| **RMSE reached at** **with no of iterations** | 231.28 200 | 317.65 215 | 167.22 440 | 208.38 1000 |
| **RMSE improved MLP (1000 it.)** **Relation to Clustering** | 227.09 +2.0% | 292.477 +7.9% | 165.641 +1.2% | 210.238 +1.9% |

Table 10.3: The RMSE for the clustering and the MLPs.

and the datum $x_j$. The activation of a neuron of the MLP is calculated by

$$f(\sum_{i=1}^{C} w_i \cdot a_i)$$

out of the activations $a_i$ of the precedent layer with $f$ being the logistical function $f(x) = \frac{1}{1-e^{-x}}$. We can plot these functions when having only two prototypes.

When plotting these functions, we see, that in figure 10.2 the function reaches 1 in the prototype. The activation function of the MLP will never reach 1, it is only approaching 1 as drawn in figure 10.3.

Additionally the slope of the membership function for the clustering depends on the distance towards the other clusters. On the other hand, the slope of the activation function is an absolute value that is fixed by the weights. We can describe the same hyperplane by defining the function with different slopes. The steeper the function, the clearer the distinction between the classes.

After having learned, the MLP performs even better than the clustering. One reason for this may be, that the clustering was performed completely without considering the travel times. The clusters were formed out of the weather data information. Then the resulting clusters were assigned to output data, that were determined by calculating the median value of the input data of

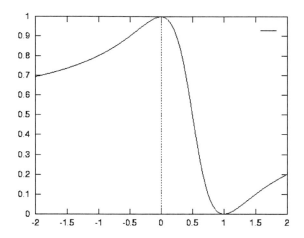

Figure 10.2: Membership degree when having two clustering prototypes

the cluster.

The MLP also considers the delay time to perform the learning algorithm and is therefore able to adapt better to the problem than the clustering algorithm.

The fact, that the MLP only improves a small amount, shows us, that the clustering has already done good work when only clustering the weather data without taking into account the delay times.

Assume that clustering only the weather data without the travel times could have resulted in an insufficient clustering. Then the learned MLP would have improved much more than it did here.

**Remark 30**

*We can give an idea of the reason why an MLP is able to improve the results of the clustering by the following considerations.*

*The fuzzy-c-means algorithm distinguishes between the clusters only by the fact, which cluster is the closest by distance. Transferring this geometrical view into an MLP leads to values of 0 and 1 for the weights for the second hidden layer.*

*During the learning process the MLP leaves these values and chooses weights somewhere between 0 and 1. The resulting MLP does not only distinguish by separating hyperplanes, but the hyperplanes are also provided with a factor of importance for the separation process.*

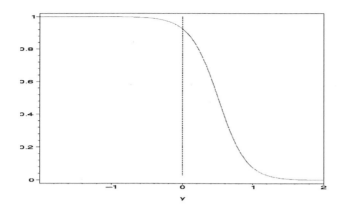

Figure 10.3: Activation function of a neuron in an MLP

|  | clouded weather | cloudless weather |
|---|---|---|
| **RMSE Clustering in 1999** | 317.63 | 214.27 |
| **RMSE 1999 data in 1998 MLP** | 349.199 | 208.383 |
| **RMSE improved MLP** **after no. of epoches** | 317.236 138 | 213.524 251 |

Table 10.4: The MLPs of 1998 are trained with 1999 data to adapt to the following year.

As the situation in air traffic control normally changes continually over a period of time, is makes sense to adapt a prediction system continously during this time. A neural network can learn while being in charge and adapt to the new situations without any need to train it completely again.

To demonstrate this we have trained the MLPs that resulted from the data of 1998 with the data of 1999. Already after about 140 resp. 250 epoches, the neural networks had reached the RMSE of those MLPs that were deducted from the clustering for 1999 and and trained with the data. The results are noted in table 10.4.

We realize that these multilayer perceptrons learn much faster when starting with the MLP for the previous year than they learn when starting with the

| Units in the hidden layer | clouded 1998 | clouded 1999 | cloudless 1998 | cloudless 1999 |
|---:|:---:|:---:|:---:|:---:|
| 5 | 221.1 | 321.5 | 170.2 | 206.5 |
| 10 | 222.4 | 316.9 | 169.5 | 210.8 |
| 25 | 230.5 | 354.4 | 172.2 | 225.0 |
| 50 | 260.2 | 365.8 | 177.5 | 234.7 |
| 100 | 285.8 | 399.7 | 176.2 | 237.8 |

Table 10.5: RSME of test dataset for MLPs trained without initialization.

| | clouded 1998 | clouded 1999 | cloudless 1998 | cloudless 1999 |
|:---|:---:|:---:|:---:|:---:|
| **Initialized MLP (training set)** | 240.1 | 346.7 | 171.2 | 216.8 |
| **Trained MLP (training set)** | 224.2 | 304.6 | 165.9 | 211.8 |
| **Trained MLP (test data)** | 230.4 | 313.0 | 171.5 | 218.7 |

Table 10.6: RSME for MLPs trained with 80% of data after initialization.

MLP deducted from this years clusters. Thus it is not necessary to cluster the data for each year and perform the transformation algorithm. It is much faster to adapt the MLP of the previous year.

This aspect is even more important as in practical use the neural network has to adapt daily. It does not make sense to wait until the end of year to withdraw information from the data.

## 10.3  Uninitialized Multilayer Perceptrons

We have seen the advantages of working with MLPs constructed from fuzzy clusters. Now the question occurs whether an MLP could perform as well without any initialization.

[87] has examined the behaviour of MLPs with different numbers of hidden neurons without initialization. The weather data were used as input, and the single output data was supposed to learn the travel time. There was only one hidden layer, and the number of units in this hidden layer was varied.

The MLPs were trained by the holdout-method. They were trained with 80% of the data and afterwards checked with the remaining 20%. To achieve representative results, this was repeated several times with different devision of the data into training- and testingset.

In table 10.5 we see the results for different sizes. We see that with high numbers of neurons in the hidden layer, the MLP starts to overadapt to the training data set and are less able to handle unknown data.

Now we used the holdout-method to train the MLP that we have constructed from the fuzzy clusters. Out networks have 27 resp. 28 units in the first hidden layer, with two more hidden layers following.

If we compare the performance of the initalized MLP with the performance of the uninitialized MLP with 25 units, we see, that we perform similar or better on the testing set than the unitilialized, although our constructed MLP has a slightly higher number ob hidden neurons that the uninitialized MLP. As the data are distributed consistently over the data space, the additional layer of the constructed MLP does not help to add structure to the solution, but represents only a number of additional hidden neurons.

This means that the effect of overadaption is less, if the MLP was already initialized close to the desired results.

## 10.4 Conclusion and Outlook

The results show us, that the MLP can improve the results of the clustering done by [87], but it also shows, that already the weather data hold the information that is necessary for good clustering results.

We have seen that the results of the clustering could be improved by using an MLP. When initializing the MLP by using the prototypes of the clusters for the construction, this leads to a fast learning MLP.

We know that the delays in arrival do not depend only on the weather but also on many other factors, but most of them do not have been well examined yet. As with different initializations the clustering algorithm finds nearly always the same clusters, we know that they are not found by accident. But the delay time of the data of one cluster is widely spread due to the influence of the other factors apart from the weather.

The results may be even more distinct for data sets with data that have less widely spread output data than our weather data.

There are not only those factors as the runway capacity, controller and pilot workload, quality of the technical infrastructure and mix of different aircraft types as well as airspace organization, that have not been analyzed here, but also the weather data is not complete. The measurements are taken only at fixed points, but the attributes can adopt quite different values at the concrete location where the aircraft passes. This is an additional reason for the widely spread data.

Examining one aspect after the other in future work will continuously lead to more and more reliable predictions.

After having constructed an MLP from the clustering results, we can further use this neural network and train it. We can adapt the MLP to current values by training it continously over time instead of doing the whole clustering again to adapt to current values.

# Chapter 11

# Conclusions and Future Perspectives

In this work we have offered some new possibilities to use rule based fuzzy systems, fuzzy clustering and neural network and presented methods to combine them.

We have demonstrated that the max-min fuzzy classification system is restricted because it decides locally on the basis of two attributes. To resolve these restrictions and to be able to solve piecewise linear separable classification problems, we examined fuzzy classifiers with the Lukasievicz-t-norm and the bounded sum.

Coming from the Lukasievicz fuzzy classification system, we have established several methods to transform different systems into each other on the basis of a geometric visualization. We have shown that the rule based fuzzy classification system as well as fuzzy clusters and multilayer perceptrons describe a classification by the means of separating hyperplanes in the data space. We use this common description to establish the procedures to transform these systems into each other.

There are different ways to make use of these transformations for applications.

When we have the resulting prototypes of a fuzzy clustering algorithm, we can transform them into a rule based classification system to take advantage of easily interpretable rules. This method is preferable to the standard method of projection, that is endowed with a loss of information.

The clustering results can also be transformed into a multilayer perceptron. This offers several possibilities. With data as in the application example of weather data, we were able to cluster the data for the first year, but did not have to repeat the clustering procedure for the following year. The neural network can be constructed for the first year, and the following years can be learned by the net, using the clustering for the initialization.

The multilayer perceptron can often perform better than the clustering when used for prediction purposes due to the fact that the MLP can assign an importance to the hyperplanes defined by the first hidden layer and does not only decide by the distance from the hyperplane.

As the example with the weather data has shown, the trained MLP can also confirm that the clustering algorithm has found reasonable prototypes. If the MLP is only capable to improve small steps, then the clustering has already been successful. Otherwise the MLP would improve the results significantly.

The last combination is that one between the rule based fuzzy classification system and the multilayer perceptron. A rule base can be set up by different methods, e.g. by realizing expert knowledge in fuzzy rules. We can transform these rules into a multilayer perceptron and improve the network by learning. This makes sense if we have training data, that contain more information than the expert can formulate, although following the experts' rules. Then the rules can be used to initialize the MLP, that is capable to draw the information out of the data set.

As the MLP usually performs very well, it is of great interest for the user to be able to understand the way the MLP solves a problem. Deriving fuzzy rules from the MLP leads us to resolving the black box characteristic of the multilayer perceptron. Although the calculation and simplification of the boolean expression represented by a given MLP bears the risk of loosing information, the ability to transform the weights of the MLP into fuzzy rules enables the expert to extract linguistic rules that are interpretable.

The different methods considered in this work all have their unique advantages. By converting them we are able to combine these advantages to profit from them while solving a problem.

As increasing air traffic results in increasing delay times, air traffic management systems aim at reducing these delays as far as possible. Their principle goal is to increase airport capacity without increasing delay times.

Examining the data of precedent years can reveal the importance of different factors on the delay times. One of these aspects is weather data and their influence on the aircrafts. Therefore we used soft computing methods on these data.

The weather data were clustered to find structures in the data set. In our example the resulting clusters could be transformed into a multilayer perceptron, using the knowledge about the geometrical structure of the data. Training the MLP provided us with further information about the data. We improved the prediction possibilities by using the neural network, and we learned that the clustering already found successfully the present structures.

# Appendix A

# The Taylor Expansion Restriction

In chapter 3 we had to assume that the fuzzy sets are continuous and have a local one-sided Taylor-expansion. This assumption was necessary to decide which of two fuzzy sets increases faster than the other.

When two functions $f_1$ and $f_2$ are given that have a Taylor expansion in $x_0$, we want to use the Taylor expansions to know which function has the greater values, when going a very small step from $x_0$ into one direction.

Therefore, we take the first term of the Taylor expansion - say teh coefficient for $(x-x_0)^n$ - that is different for the two functions. Without loss of generality we have $f_1^{(n)}(x_0) > f_2^{(n)}(x_0)$ (and with this the $n^{th}$ coefficient of the Taylor expansion of $f_1$ is greater than that one of $f_2$), while for $i \in \{1, \cdots, n\}$ we have $f_1^{(i)}(x_0) = f_2^{(i)}(x_0)$. Then the values of $f_1$ are greater than those for $f_2$, when $x > x_0$, and the other way round for $x < x_0$ in the neighborhood $\mathcal{N}_\varepsilon(x_0)$ of $x_0$, as the following argument will show.

**Lemma 7**
*Assume $f : \mathbb{R} \to \mathbb{R}$ is twice differentiable in a neighborhood of $x_0$ and let $t$ be the tangent to $f$ at point $x_0$. $t$ has the slope $m_t = f'(x_0)$. Let $g$ and $h$ be straight lines with $g(x_0) = h(x_0) = f(x_0)$ and with $g$ having slope $m_g > m_t$ and $h$ having slope $m_h < m_t$.*

*Then there is an $\varepsilon > 0$ so that in $\mathcal{N}_\varepsilon(x_0)$ $f$ lies between $g$ and $h$. This means:*

$$\forall x \in \mathcal{N}_\varepsilon(x_0), x < x_0 : \quad g(x) < f(x) < h(x) \quad \text{and}$$
$$\forall x \in \mathcal{N}_\varepsilon(x_0), x > x_0 : \quad g(x) > f(x) > h(x).$$

**Proof:**   We can write $t(x) = f(x_0) + f'(x_0) \cdot (x - x_0)$ and with this

$$f(x) = t(x) + r(x)(x - x_0)^2$$

with $r(x_0) = 0$ and $r$ being a continuous function. $r$ being continuous means

$$\forall\, \delta > 0 \; \exists\, \varepsilon > 0 : (|x - x_0| < \varepsilon \Rightarrow |r(x) - r(x_0)| = |r(x)| < \delta).$$

Considering $x > x_0$ with $x - x_0 \leq \min\{\epsilon, 1\}$ we choose $\delta := m_g - f'(x_0)$ and obtain

$$g(x) - f(x) = \underbrace{\underbrace{(m_g - f'(x_0)}_{=\delta>0} - \underbrace{r(x) \cdot (x - x_0))}_{<\delta\cdot 1}}_{>0} \cdot \underbrace{(x - x_0)}_{>0} > 0,$$

because $r$ is continuous. So we have $g(x) > f(x)$ for $x > x_0$, $x \in \mathcal{N}_\varepsilon(x_0)$. By the same argument we obtain $f(x) > h(x)$ for $x > x_0$ and $h(x) > f(x) > g(x)$ for $x < x_0$ within $\mathcal{N}_\varepsilon(x_0)$.   $\square$

**Lemma 8**

Assume $f_1$ and $f_2$ are twice differentiable in a neighborhood of $x_0$ and let $f_1(x_0) = f_2(x_0)$, but $f_1'(x_0) > f_2'(x_0)$. Then there are a straight line $g$ with $g(x_0) = f_1(x_0) = f_2(x_0)$ and an $\varepsilon > 0$ so that $g$ lies between $f_1$ and $f_2$ within the neighborhood $\mathcal{N}_\varepsilon(x_0)$ of $x_0$. This means that

$$\begin{aligned}
f_1(x) < g(x) < f_2(x) \quad &\text{for } x < x_0 \quad \text{and} \\
f_1(x) > g(x) > f_2(x) \quad &\text{for } x > x_0.
\end{aligned}$$

**Proof:**   Define $g$ by

$$g(x) = f_1(x_0) + \frac{f_1'(x_0) + f_2'(x_0)}{2}(x - x_0)$$

with slope $m_g = \frac{f_1'(x_0)+f_2'(x_0)}{2}$. considering two straight lines $h_1$ and $h_2$ with $h_1(x_0) = h_2(x_0) = g(x_0)$ with slopes $m_{h_1} > f_1'(x_0)$ and $m_{h_2} < f_2'(x_0)$, we can apply Lemma 7 to show that there is an $\varepsilon$ so that $g$ lies between $f_1$ and $f_2$ in $\mathcal{N}_\varepsilon(x_0)$.   $\square$

**Lemma 9**

Assume, the functions $f_1$ and $f_2$ are $(n + 2)$ times differentiable in a neighborhood of $x_0$ and let $f_1^{(i)}(x_0) = f_2^{(i)}(x_0)$ for $i = 0, \cdots, n$, but $f_1^{(n+1)}(x_0) > f_2^{(n+1)}(x_0)$. Then there is an $\varepsilon > 0$ so that in $\mathcal{N}_\varepsilon(x_0)$ we have $f_1(x) < f_2(x)$ for $x < x_0$ and $f_1(x) > f_2(x)$ for $x > x_0$.

**Proof:** We give a proof by induction:

Beginning of induction ($n = 0$): Let $f_1(x_0) = f_2(x_0)$ and $f_1'(x_0) > f_2'(x_0)$. Because of lemma 8 we can put a straight line between $f_1$ and $f_2$. So we have $f_1(x) > f_2(x)$ for $x > x_0$, $x \in N_\varepsilon(x_0)$, and the other way round for $x < x_0$, $x \in N_\varepsilon(x_0)$.

Induction hypothesis: When we have $\tilde{f}_1^{(i)}(x_0) = \tilde{f}_2^{(i)}(x_0)$ for $i = 0, ..., n-1$ and $\tilde{f}_1^{(n)}(x_0) > \tilde{f}_2^{(n)}(x_0)$, then there is an $\varepsilon > 0$ so that $\tilde{f}_1(x) > \tilde{f}_2(x)$ for $x > x_0$, $x \in N_\varepsilon(x_0)$, and the other way round for $x < x_0$, $x \in N_\varepsilon(x_0)$.

Induction step: We have $f_1^{(i)}(x_0) = f_2^{(i)}(x_0)$ for $i = 1, ..., n$ and $f_1^{(n+1)}(x_0) > f_2^{(n+1)}(x_0)$. When defining $\tilde{f}_1 := f_1'$ and $\tilde{f}_2 := f_2'$ we can use the hypothesis and calculate for $x = x_0 + \delta$, $0 < \delta < \varepsilon$:

$$
\begin{aligned}
f_1(x) - f_2(x) &= f_1(x_0 + \delta) - f_2(x_0 + \delta) \\
&= \int_0^\delta (f_1'(x_0 + t) - f_2'(x_0 + t)) dt \\
&= \int_0^\delta \underbrace{(\tilde{f}_1(x_0 + t) - \tilde{f}_2(x_0 + t))}_{\geq 0} dt > 0,
\end{aligned}
$$

because we have $\tilde{f}_1(x_0 + t) > \tilde{f}_2(x_0 + t)$ for $0 < t < \varepsilon$. Therefore, we obtain $f_1(x) > f_2(x)$ for $x > x_0$, $x \in N_\varepsilon(x_0)$. The same can be carried out for $x < x_0$, $x \in N_\varepsilon(x_0)$. $\qquad\square$

**Corollary 10**

Let $f_1$ and $f_2$ be two functions so that $f_i(x) = \sum_{k=0}^\infty a_k^{(i)}(x - x_o)^k$ for $x_0 \leq x \leq b$ and let $a_0^{(1)} = a_o^{(2)}$, i.e. $f_1(x_0) = f_2(x_0)$, then either

1. $\forall x \in [x_0, b] : f1(x) = f_2(x)$ or

2. $\exists \varepsilon > 0$ so that either $\forall x$ with $0 < x - x_0 < \varepsilon$: $f_1(x) < f_2(x)$ or $\forall x$ with $0 < x - x_0 < \varepsilon$: $f_1(x) < f_2(x)$

holds.

**Proof:** If $a_k^{(1)} = a_k^{(2)}$ for all $k \in N$, then $f_1(x) = f_2(x)$ for all $x \in [x_0, b]$. Otherwise Lemma 9 is applicable. $\qquad\square$

**Corollary 11**
Let $f_1$ and $f_2$ be two functions so that $f_i(x) = \sum_{k=0}^{\infty} a_k^{(i)}(x-x_o)^k$ for $b \leq x \leq x_0$ and let $a_0^{(1)} = a_o^{(2)}$, i.e. $f_1(x_0) = f_2(x_0)$, then either

1. $\forall x \in [b, x_0] : f1(x) = f_2(x)$ or

2. $\exists \varepsilon > 0$ so that either $\forall x$ with $0 < x_0 - x < \varepsilon$: $f_1(x) < f_2(x)$ or $\forall x$ with $0 < x_0 - x < \varepsilon$: $f_1(x) < f_2(x)$

holds.

**Proof:** If $a_k^{(1)} = a_k^{(2)}$ for all $k \in N$, then $f_1(x) = f_2(x)$ for all $x \in [x_0, b]$. Otherwise Lemma 9 is applicable. □

# Appendix B

# The algorithms

In this chapter we have collected the algorithm that appear in the chapters of this thesis. In the chapters we described the agorithms in a heuristical way. In this appendix they will be given more concrete to make it easier to transform them into computer programms.

## B.1 Calculating the Rules with the Łukasiewicz-t-Norm

In chapter 4 we explained how to calculate the rules for a fuzzy classification system based on the Łukasiewicz-t-norm. Here we will summarize the calculations for easier usage.

$[a_1, b_1] \times [a_2, b_2]$ is the rectangle. To keep the notation simple, we describe the fuzzy sets as in chapter 4. A data structure for the fuzzy sets should consist of the boundaries $a_i$ and $b_i$ of the interval, in which the fuzzy set has a membership degree greater than zero, of the slope and the abscissa. A rule consists of $n$ such fuzzy sets – one for each dimension – and of an integer to note the class the rule is firing for.

**Algorithm 7 (Rectangle of Type 0)**

$void\ rectangle\_type\_0(double\ a_1,\ a_2,\ b_1,\ b_2,\ int\ class)$
$\{$

$\quad rule\ R;$

$$R.\mu_R^{(1)}(x) := \begin{cases} \max\{0, \min\{1, \frac{x-a_1}{b_1-a_1}\}\} & if\ x \in [a_1, b_1]; \\ 0 & otherwise; \end{cases}$$

$$R.\mu_R^{(2)}(y) := \begin{cases} \max\{0, \min\{1, \frac{y_2-a_2}{b_2-a_2}\}\} & \text{if } y \in [a_2, b_2]; \\ 0 & \text{otherwise}; \end{cases}$$

$\quad$ *add_to_R(R, class)*

}

$\diamond$

In a rectangle of type 1 we need the two points $(x_1, y_1), (x_2, y_2)$ where the separation line meets the boundaries of the rectangle.

## Algorithm 8 (Rectangle of Type 1)

$\quad$ *void rectangle_type_1(double $a_1$, $a_2$, $b_1$, $b_2$, $x_1$, $y_1$, $x_2$, $y_2$,*
$\qquad$ *int class_of_($a_1, b_1$), class_of_($a_2, b_2$))*

{

$\quad$ *rule* $R^1$, $R^2$;

$\quad$ *if* $(x_1 \neq x_2)$

$\qquad$ *then*

$\qquad$ {

$$\mu_{R^1}^{(1)}(x) := \begin{cases} \max\{0, \min\{1, \frac{x-x_1}{b_1-x_1}\}\} & \text{if } x \in [a_1, b_1]; \\ 0 & \text{otherwise}; \end{cases}$$

$$\mu_{R^2}^{(1)}(x) := \begin{cases} \max\{0, \min\{1, \frac{x_2-x}{x_2-x_1}\}\} & \text{if } x \in [a_1, b_1] \\ 0 & \text{otherwise} \end{cases} ;$$

$$\mu_{R^1}^{(2)}(y) := \begin{cases} \max\{0, \min\{1, \frac{y_2-y}{y_2-y_1}\}\} & \text{if } y \in [a_2, b_2] \\ 0 & \text{otherwise} \end{cases}$$

$$\mu_{R^2}^{(2)}(y) := \begin{cases} \max\{0, \min\{1, \frac{y-y_1}{y_2-y_1}\}\} & \text{if } y \in [a_2, b_2] \\ 0 & \text{otherwise} \end{cases} ;$$

$\qquad\quad$ *add_to_R($R^1$, class_of_($a_1, b_1$));*
$\qquad\quad$ *add_to_R($R^2$, class_of_($a_2, b_2$));*

$\qquad$ }

$\qquad$ *else*

$\qquad$ {

$$\mu_{R^1}^{(1)}(x) := \begin{cases} \max\{0, \min\{1, \frac{x_2-x}{b_1-a_1}\}\} & \text{if } x \in [a_1, b_1] \\ 0 & \text{otherwise} \end{cases} ;$$

$$\mu_{R^2}^{(1)}(x) := \begin{cases} \max\{0, \min\{1, \frac{b_1-x}{b_1-a_1}\}\} & \text{if } x \in [a_1, b_1] \\ 0 & \text{otherwise} \end{cases} ;$$

$$\mu_{R^1}^{(2)}(y) := \begin{cases} \max\{0, \min\{1, \frac{b_2-y}{b_2-b_1}\}\} & \text{if } y \in [a_2, b_2] \\ 0 & \text{otherwise} \end{cases}$$

$$\mu_{R^2}^{(2)}(y) := \mu_{R^1}^{(2)}(y)$$

```
            add_to_R(R¹, class_of_(a₁, b₁));
            add_to_R(R², class_of_(a₂, b₂));
        }
}
```

◇

In chapter 4 we described two possibilities for the rectangle with the acute angle.

For the first possibility, we have to choose the size of the rectangle containing the acute angle in a way that we can except the misclassification $\alpha$ but still do not have too many rectangles. Then the first rectangle is calcuated, and the other rectangles follow depending on the boundaries of the first one.

**Algorithm 9 (Rectangle of Type 3, neglecting small rectangle)**

*void rectangle_type_2(double $a_1$, double $a_2$, double $b_1$, double $b_2$, double $c_2$,*
*        int class_inside, int class_outside, double alpha)*

```
{
```
> *double* $\delta_1 := \sqrt{\alpha \cdot \frac{b_1 - a_1}{b_2 - a_2}}$;
>
> *double* $\delta_2 := \frac{\alpha}{\delta_1}$;
>
> *rectangle_type_0($a_1$, $a_2$, $a_1 + \delta_1$, $a_2 + \delta_2$, class_inside);*
>
> *rectangle_type_0($a_1$, $a_2 + \delta_2$, $a_1 + \delta_1$, $b_2$, class_outside);*
>
> *rectangle_type_1($a_1 + \delta_1$, $a_2$, $b_1$, $a_2 + \delta_2$, $a_1 + \delta_1$, $\frac{\delta_1 \cdot (c_2 - a_2)}{(b_1 - a_1)^2}$,*
>
> > $a_1 + (a_2 + \delta_2) \cdot \frac{(b_1 - a_1)^2}{c_2 - a_2}$, $a_2 + \delta_2$, *class_outside, class_inside);*
>
> $c_1 := a_1 + (a_2 + \delta_2) \cdot \frac{(b_1 - a_1)^2}{c_2 - a_2}$;
>
> $a_1 := a_1 + \delta_1$;
>
> $a_2 := a_2 + \delta_2$;
>
> $m_1 := \frac{b_2 - a_2}{b_1 - a_1}$;
>
> $m_2 := \frac{c_2 - a_2}{b_1 - a_1}$;
>
> *while ($c_2 > a_2 + m_1 \cdot (c_1 - a_1)$)*
> ```
> {
> ```
> > *rectangle_type_1($a_1$, $a_2$, $c_1$, $b_2$, $a_1$, $b_1$, $c_1$, $a_2 + m_1 \cdot (c_1 - a_1)$,*
> > > *class_outside, class_inside);*
> >
> > *rectangle_type_1($c_1$, $a_2$, $b_1$, $a_2 + m_1 \cdot (c_1 - a_1)$, $c_1$, $a_2$,*
> > > $c_1 + \frac{m_1}{m_2} \cdot (c_1 - a_1)$, $a_2 + m_1 \cdot (c_1 - a_1)$,
> > > *class_inside, class_outside);*
> >
> > $a_1 := c_1$;
> >
> > $a_2 := a_2 + m_1 \cdot (c_1 - a_1)$;
> ```
> }
> ```
> *rectangle_type_1($a_1$, $a_2$, $c_1$, $b_2$, $a_1$, $b_1$, $c_1$, $a_2 + m_1 \cdot (c_1 - a_1)$,*

$$class\_outside,\ class\_inside);$$
$$rectangle\_type\_1(c_1,\ a_2,\ b_1,\ a_2 + m_1 \cdot (c_1 - a_1),\ c_1,\ a_2,\ b_1,$$
$$a_2 + m_2 \cdot (b_1 - a_1),\ class\_inside,\ class\_outside);$$
$$rectangle\_type\_1(c_1,\ a_2 + m_1 \cdot (c_1 - a_1),\ b_1,\ b_2,\ c_1,\ c_2,\ b_1,\ b_2,$$
$$class\_outside,\ class\_inside);$$

$\diamond$

For the second possibility the rules can be calculated straight foreward. The resulting rules give an exact separation, but they do not all reach membership degree 1.

### Algorithm 10 (Rectangle of Type 3, exact solution)

$$void\ rectangle\_type\_2(double\ a_1,\ double\ a_2,\ double\ b_1,\ double\ b_2,\ double\ c_2,$$
$$int\ class\_inside,\ int\ class\_outside)$$
{

    $rule\ R;$

$$\mu_R^{(1)}(x) := \begin{cases} \max\{0, \min\{1, 1 - \frac{x-a_1}{2(b_1-a_1)}\}\} & if\ x \in [a_1, b_1]; \\ 0 & otherwise; \end{cases}$$

$$\mu_R^{(2)}(y) := \begin{cases} \max\{0, \min\{1, \frac{y-a_2}{2(b_2-a_2)}\}\} & if\ y \in [a_2, b_2]; \\ 0 & otherwise; \end{cases}$$

    $add\_to\_\mathcal{R}(R, class\_inside\ );$

$$\mu_R^{(1)}(x) := \begin{cases} \max\{0, \min\{1, \frac{x-a_1}{2(b_1-a_1)}\}\} & if\ x \in [a_1, b_1]; \\ 0 & otherwise; \end{cases}$$

$$\mu_R^{(2)}(y) := \begin{cases} \max\{0, \min\{1, 1 - \frac{y-a_2}{2(b_2-a_2)}\}\} & if\ y \in [a_2, b_2]; \\ 0 & otherwise; \end{cases}$$

    $add\_to\_\mathcal{R}(R, class\_outside\ );$

    $double\ \beta := \frac{c_2-a_2}{b_2-a_2};$

$$\mu_R^{(1)}(x) := \begin{cases} \max\{0, \min\{1, \frac{1+\beta}{2} \cdot \frac{x-a_1}{b_1-a_1}\}\} & if\ x \in [a_1, b_1]; \\ 0 & otherwise; \end{cases}$$

$$\mu_R^{(2)}(y) := \begin{cases} \max\{0, \min\{1, 1 - \frac{y-a_2}{b_2-a_2}\}\} & if\ y \in [a_2, b_2]; \\ 0 & otherwise; \end{cases}$$

    $add\_to\_\mathcal{R}(R, class\_outside\ );$

}

$\diamond$

# B.2 Classification Rules from Fuzzy Clusters

In chapter 7 we transformed fuzzy clusters into a fuzzy classification system. In this appendix the algorithms for processing the transformation can be found.

The first steps are the calculation of the intersection point $P_s$, the center of gravity $P_g$ and the line $l$ that passes by $P_s$ and $P_g$. This line $l$ is defined by its direction $P_g - P_s$ and the point $P_s$ that it passes.

**Algorithm 11 (Initialization)**

*Void Initialize(Cuboid $[a_1, b_1] \times \cdots \times [a_n, b_n]$, HyperplaneList H)*
*{*
    *Vector $P_s := H[1] \cap \ldots \cap H[h]$;*
    *HyperlineList l;*
    *for all $i = 1, \ldots, h$*
        *for all $j = 1, \ldots, h$ with $j \neq i$*
        *$l_{ij} := H[i] \cap H[j]$;*
        *if ($l_i j$ on border of section)*
        *then*
            *$l.Add(l_{ij})$;*
    *double $\lambda := length (l)$;*
    *VectorArray X;*
    *X.Set_Length():= $\lambda$;*
    *for all $i = 1, \ldots, \lambda$*
        *$X[i] := \lambda \cap$ Boundaries of the cuboid;*
    *Vector $P_g := P_s + \frac{1}{\lambda} \cdot \sum_{i=1}^{\lambda} x_i$;*
    *Line $l : x = P_s + \alpha(P_g - P_s), \alpha \in \mathbb{R}$;*
*}*

$$\diamond$$

The next step is the construction of the auxiliary planes as described in section 7.1.3 on page 92. The following calculations have to be done for each hyperplane $H_i$ separately.

We want to calculate the vector $y_{H_i}$ with

$$y_{H_i} = \alpha \cdot n_{H_i} + \beta \cdot (P_g - P_s), \qquad (\alpha, \beta \in \mathbb{R}). \qquad (B.1)$$

As the length of $y_{H_i}$ can be freely chosen, we set $\alpha := 1$. Because of the fact that $y_{H_i}$ has to belong to $H_i$, we can write

$$y_{H_i} \cdot n_{H_i} + d_{H_i} = 0. \qquad (B.2)$$

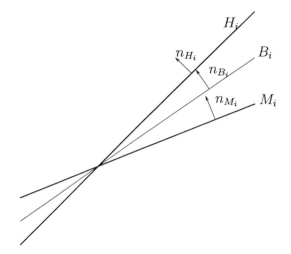

Figure B.1: The construction of the normalvector of $B_i$.

Now we insert (B.1) in (B.2) and achieve

$$\alpha \cdot n_{H_i} \cdot n_{H_i} + \beta(P_g - P_s) \cdot n_{H_i} + d = 0 \Leftrightarrow \beta = \frac{-1 - d}{(P_g - P_s) \cdot n_{H_i}}.$$

Then $y_{H_i} := n_{H_i} + \beta \cdot (P_g - P_s)$ is a solution for (B.1) and (B.2).
Now we take a basis of $H_i$. As $y_{H_i}$ belongs to $H_i$ we can start with $y_{H_i}$ to construct a orthogonal basis for $H_i$ including $y_{H_i}$ by using the procedure of Gram-Schmidt. When changing $y_{H_i}$ in this basis into $(P_g - P_s)$ we achieve a basis for the hyperplane $M_i$ and can calculate the normalvector $n_{M_i}$ of $M_i$.
Choosing a point $P_{H_i} \in H_i$ helps us to see whether $n_{M_i}$ is pointing into the direction of $H_i$ or into the other direction. If $p_{H_i} = n_{M_i} \cdot P_{H_i} + d_{M_i} > 0$, then $n_{M_i}$ is pointing into the direction of $H_i$. The same considerations hold for a point $p_{M_i}$.
Figure B.1 shows how this looks like if $n_{M_i}$ points towards $H_i$ and $n_{H_i}$ points away from $M_i$. Otherwise we can adapt the procedure by changing the direction of the normalvectors. This results in the formula

$$n_{B_i} := \frac{\operatorname{sgn}(p_{H_i}) \cdot n_{M_i} - \operatorname{sgn}(p_{M_i}) \cdot n_{H_i}}{\left| \frac{\operatorname{sgn}(p_{H_i}) \cdot n_{M_i} - \operatorname{sgn}(p_{M_i}) \cdot n_{H_i}}{\big|} \right.}.$$

As $P_s \in B_i$, we can calculate $d_{B_i} := -P_s \cdot n_{B_i}$.

**Algorithm 12 (Construction of the Auxiliary Planes)**

*Planes Auxiliary_Planes(Hyperplane H, $P_g$, $P_s$)*
{

    *double* $\beta := \frac{-1-d}{(P_g - P_s) \cdot n_{H_i}}$;

    $y_{H_i} := n_{H_i} + \beta \cdot (P_g - P_s)$;

    *VectorArray HBasis, MBasis, BBasis;*

    *HBasis := GramSchmidt* $(H_i, y_H)$;

    *MBasis := HBasis;*

    *MBasis.Remove*$(y_H)$;

    *MBasis.Add*$(P_g - P_s)$;

    $n_{M_i} := $ *Normalvector(MBasis)*;

    $d_{M_i} := $ *Distance(MBasis)*;

    *Vector P1, P2;*

    *double p1, p2;*

    *Choose point p1* $\in H$ *and* $p_2 \in M$;

    $p1 = n_{M_i} \cdot p1 + d_M$;

    $p2 = n_{M_i} \cdot p1 + d_M$;

    $n_{B_i} := \text{sgn}(p1) \cdot n_{M_i} - \text{sgn}(p) \cdot n_{H_i}$;

    *normalize* $n_{B_i}$;

    $d_{B_i} := -P_s \cdot n_{B_i}$;

    *return (M ,B);*

}

$\Diamond$

We have calculated the auxiliary planes $M_i$ and $B_i$. With these hyperplanes we can calculate the rules.

The function Choose_P calculates the corner of the examined cuboid that belongs to a hyperplane, i.e. that is the farthest corner outside the cuboid. The principle behind the algorithm to determine this corner is simple:

We start with a point inside the classified section. One possibility is $P_g$. Then we check one coordinate $x_i$ after the other. We choose either $a_i$ or $b_i$, depending on the fact, which one leads us closer to the hyperplane $H_i$ or even lets us get to the other side of $H_i$, i.e. outside the classified section.

Then we continue by choosing $a_i$ or $b_i$ depending on the fact which one takes us further away from $H_i$ without passing it again and getting inside the section again.

We determine the rule $\mu_{R_{M_i}}$ by a straight forward calculation as explained in section 7.1.3.

## Algorithm 13 (Construction of the Rule $R_{M_i}$)

*Point Choose_P(Point $P_g$, Hyperplane $H_i$, Cuboid)*
{
    Point $C := P_g$
    double $d1$, $d2$;
    int $j := 0$;

    $d1 := n_{M_i} \cdot P_g + d_{M_i}$;
    $d2 := d1$;
    while $(j < n$ and $d1 \cdot d2 > 0)$
        $C[j] := a_j$
        $d2 := n_{M_i} \cdot P_g + d_{M_i}$;
        if $(d1 \cdot d2 > 0)$
            if$(|d2| > |d1|)$
                $C[j] := b_j$;
                $d2 := n_{M_i} \cdot P_g + d_{M_i}$;
        $j := j + 1$;
    while$(j < n)$
        $C[j] := a_j$;
        $d2 := n_{M_i} \cdot P_g + d_{M_i}$;
        if$(|d2| > |d1|)$
            $C[j] := b_j$;
        else
            if $(|d2| < |d1|)$
                $C[j] := b_j$;
        return$(C)$;
}

*Rule Calculate_Rule(Point P, Hyperplane M, H, $[a_1, b_1] \times \cdots \times [a_n, b_n]$,*
                *int class_inside)*
{
    double $\gamma := -(d_{M_i} + \sum_{t=1}^{n}(n_{M_i}[t] \cdot P[t]))^{-1}$;
    for all $j = 1, \ldots, n$
        $\alpha := \gamma \cdot n_{M_i}[j]$
        if $(P[j] = a_j)$
        then
            $\mu_{R_{M_i}}^{(j)}(x_j) := 1 - n_{M_i} \cdot (x_j - P[j])$;

*else*

$$\mu_{R_{M_i}}^{(i)}(x_i) := 1 - n_{M_i}[j] \cdot (P[j] - x_j);$$

*return* $(\mu_{R_{M_i}})$;

}

*void MRule( Hyperplane* $M_i$, $H_i$, *Cuboid* $[a_1, b_1] \times \cdots \times [a_n, b_n]$,
    *int class_inside, class_outside)*

{

   *Point P;*

   *Rule* $R_{M_i}$;

   *P=Choose_P();*

   $R_{M_i}$ :=*Calculate_Rule(P,* $M_i$,$H_i$, $[a_1, b_1] \times \cdots \times [a_n, b_n]$, *class_inside);*

   *Add_to_R* $(R_{M_i},$ *class_inside);*

}

$\diamond$

The construction of the rule $\mu_{R_{B_i}}$ is nearly the same, but we use a different auxiliary point instead of the corner.

$P_i'$ has to belong to the hyperplane that is orthogonal to $M_i$, $B_i$ and $H_i$ and that includes $P_i$. For the construction of $P_i'$ we use a point $S_{P_i}$ that has to be the point, where $H_i$ intersects with the line

$$l' : x = P_i + \alpha \cdot n_{M_i}, \alpha \in \mathbb{R}. \tag{B.3}$$

As $H_i$ is defined by $x \cdot n_{H_i} + d_{H_i} = 0$, we can insert (B.3) and achieve

$$P_i \cdot n_{H_i} + \alpha \cdot n_{M_i} \cdot n_{H_i} + d_{H_i} = 0.$$

The firing degree of the rules has to be the same on $H_i$, and because of $S_{P_i} \in H_i$ also in $S_{P_i}$. The equation

$$\frac{\text{dist}(P_i, M_i)}{\text{dist}(S_{P_i}, M_i)} = \frac{\text{dist}(P_i', B_i)}{\text{dist}(S_{P_i}, B_i)}$$

(with the distance $\text{dist}(P_i, H) = P_i \cdot n_{H_i} + d_H$ if a hyperplane $H$ is described by the normal form $P_i \cdot n_{H_i} + d_H = 0$) holds because the firing degree of the rules increases linearly.

Equation 7.6 demonstrated the derivation of the formula for $P_i'$.

**Algorithm 14 (Construction of the Rule $R_{B_i}$)**

*Point Choose_P'(Point $P_i$, Hyperplane $M_i$, $B_i$, $H_i$, Cuboid)*
{

$$S := P_i - n_{H_i} \cdot \frac{P_i \cdot n_{H_i} + d_{H_i}}{n_{M_i} \cdot n_{H_i}};$$

$$\beta := \frac{(S \cdot n_{B_i} + n_{B_i}) \cdot (P_i \cdot n_{M_i} + n_{M_i})}{S \cdot n_{M_i} + n_{M_i}} - d_{B_i};$$

$$P_i' := S + (n_{B_i} \cdot (\beta - S \cdot n_{B_i}));$$

$$\text{return } (P_i');$$

}

*void BRule()*
{

Point $P' :=$ Choose_P'(Point $P$, Hyperplane $M_i$, $B_i$, $H_i$, Cuboid);

Rule $R_{B_i} :=$ Calculate_Rule($P_i'$, $B_i$, $[a_1, b_1] \times \cdots \times [a_n, b_n]$)

Add_to_$\mathcal{R}$ ($\mu_{R_{B_i}}$, class_outside);

}

$\diamondsuit$

Finally the rules have to be scaled as described in section 7.1.3. The scaling factors $s$ is calculated out of the

**Algorithm 15 (Scaling the Rules)**

*ListofRules Scaling($R_{M_1}, \ldots, R_{M_h}, R_{B_1}, \ldots, R_{B_h}$)*
{

    for all $i = 1, \ldots, h$

        for all $j = 1, \ldots, h$ with $j \neq i$

            choose a point $A \in H_i \cap H_j$;

$$s_{ij} := \frac{\mu_{R_{M_i}}(a_1, \ldots, a_n)}{\mu_{R_{M_j}}(a_1, \ldots, a_n)};$$

    determine $p$ by $s := \max_{i,j}\{s_{ij}\} = s_{pq}$;

    for all rules $R_{M_i}$, $i = 1, \ldots, h$

        for all $t = 1, \ldots, n$

            $\gamma_t = n_{M_i}[t]$

$$\bar{\gamma} := -(c + \sum_{t=1}^{n} \gamma_t(p_t - \tau \cdot (1 - s_{pj}^{-1}) \cdot n_{M_i}^{(t)}))^{-1};$$

$$\gamma = -(c + \sum_{t=1}^{n} \gamma_t \cdot p_t);$$

for all $t=1,\ldots,n$

if $(\mu_{R_{M_i}}(x_t) = \max\{0, \min\{1, 1 - \alpha_t(x_t - p_t)\}\})$

then

$$\bar{\mu}_{R_{M_i}}^{(t)}(x_t) = 1 - \alpha_t \cdot \frac{\bar{\gamma}}{\gamma} \cdot (x_t - p_t + \tau \cdot (1 - \frac{1}{s_{pj}}) \cdot n_{M_i}[t]);$$

for all rules $R_{B_i}$, $i = 1, \ldots, h$

for all $t = 1, \ldots, n$

$\gamma_t = n_{B_i}[t];$

$\bar{\gamma} := -(c + \sum_{t=1}^{n} \gamma_t(p_t - \tau \cdot (1 - s_{pj}^{-1}) \cdot n_{B_i}^{(t)}))^{-1};$

$\gamma = -(c + \sum_{t=1}^{n} \gamma_t \cdot p_t);$

for all $t = 1, \ldots, n$

if $(\mu_{R_{B_i}}(x_t) = \max\{0, \min\{1, 1 - \alpha_t(x_t - p_t)\}\})$

then

$$\bar{\mu}_{R_{B_i}}^{(t)}(x_t) = 1 - \alpha_t \cdot \frac{\bar{\gamma}}{\gamma} \cdot (x_t - p_t + \tau \cdot (1 - \frac{1}{s_{pj}}) \cdot n_{B_i}[t]);$$

return *(ListofRules)* }

$\diamond$

The rules that are returned by the scaling algorithm are the rules that give the exact classification defined ba tha hyperplanes.

# B.3 Practical Algorithm for Calculating a Multilayer Perceptron from Fuzzy Clusters

In the previous section, we have seen the steps needed to calculate fuzzy rules from fuzzy clusters. In this section, we explain, how to determine an MLP that is calculating the same classification as a given fuzzy clustering. We show how the implementation was performed for the application in chapter 10.

### Algorithm 16 (Calculating an MLP from Fuzzy Clusters)

void *MLP_from_Clusters* (*VectorArray Dataspace* $= [a_1, b_1] \times \cdots \times [a_n, b_n]$;
*List_of_vectors Prototypes* $= \{c_1, \ldots, c_c\}$)
{

Vector $P_s$;
*DoubleVectorArray Hyperplanes_to_examine*;
*BoolVectorArray Possible_Combinations*;

Hyperplanes $H := \{H_{ij} | i, j \in \{1, \ldots, c,\ \text{normal vector pointing of } H_{ij}$
from $c_j$ to $c_i\}$;
Mark all $H_{ij} \in H$ irrelevant;
for all $i := 1, \ldots, c$
  $k :=$ Init_Hyperplanes_to_examine($i$, * Hyperplanes_to_examine);
  Initialize_Possible_Combinations($k + 2 \cdot n$);
  for all $H_{ij}, j > i$, $H_{ij}$ not yet marked relevant
    for all $l := 1, \ldots, \binom{(z+2 \cdot n)}{n}$
      BoolVector $x_b := l^{th}$ vector of Possible_Combinations;
      if $(x_b(H_i j) = 1)$
      then
        if (for all $r = 1, \ldots, n$: if $x_b[a_r] = 1$ then $x_b[b_r] = 0$
          and vice versa)
        then
          $P_s := \bigcap \{H_{st} \in H, x_B[H_{st}] = 1\}$
           $\bigcap \{B \text{ boundary of Dataspace}, x_B[B] = 1\}$;
          if ($Relevant(P_s)$)
          then
            for all $H_{st}$ with $x_B[H_{st}] = 1$
              mark $H_{st}$ to be relevant;
  Construct_MLP(List_of_Relevant_Hyperplanes);
}

                         $\diamond$

The hyperplanes between the current cluster $c_i$ and each of the previous clusters $c_j, j = 1, \ldots, i-1$ have already been examined, and we only have to take them into account if they are relevant. The other hyperplanes $H_{ij}, j = i + 1, \ldots, c$ still have to be checked.

## Algorithm 17 (Init_Hyperplanes_to_examine)

int Initialize_Hyperplanes_to_examine(int $i$)
{
  for j=1,...,i-1
    if ($H_{ij}$ relevant)
    then
      add $H_{ij}$ to Hyperplanes_to_examine;
  for $j = i + 1, \ldots, c$
    add $H_{ij}$ to Hyperplanes_to_examine;
  return(number of hyperplanes in Hyperplanes_to_examine);
}

                         $\diamond$

*Hyperplanes_to_examine* holds a list of hyperplanes. We have to check all combinations of these hyperplanes with each other and with the boundaries of the data space. We use the boolean auxiliary VectorArray *Possible_Combinations* with $z + 2 \cdot n$ columns. The first $z$ columns represent the hyperplanes that have to be examined, the following $n$ columns represent the boundaries of $[a_1, b_1] \times \cdots [a_n, b_n]$ at $a_i$ and the last $n$ columns at $b_i$.

In each Vector we note a possible combination of 0 and 1, and each Vector represents a possible intersection point. If there is a 1 noted, then the hyperplane will be taken into account, if there is a 0, then we neglect it.

In the $n$-dimensional space, we can determine an intersection point out of $n$ hyperplanes. Therefore there are $\binom{(z+2 \cdot n)}{n}$ possible combinations, with $z$ being the number of hyperplanes that have to be examined, so that *Possible_Combinations* has $\binom{(z+2 \cdot n)}{n}$ rows.

### Algorithm 18 (Initialize Possible Combinations)

*Void Initialize_Possible_Combinations(int length)*
{
    int $k := 0$;
    *BoolVector* $x_b$;

    for $l = 1, \ldots, 2^{length}$
        $x_b := $ *digital representation of l*;
        if ($x_b$ *includes exactly n digits 1*)
        then
            *Possible_Combinations*[$i$] := $x_b$;
            $k + +$;
}

$\diamond$

When we have determined the intersection point, we still have to check whether this point is relevant. If it is situated outside the data space, the point is not relevant.

As a hyperplane has equal distance to both defining prototypes, the intersection point $P_s$ is equally close to all the prototypes that defined one of the intersecting hyperplanes. If there is another prototype, that is closer to the intersection point, then $P_s$ belongs to this cluster, and not to those that define the point. Therefore $P_s$ is not relevant, if there is a closer prototype.

**Algorithm 19 (Checking relevance of intersection point)**

> bool Relevant(Vector $P_s$)
> {
>    if ($P_s$ outside data space)
>    then
>        return(false);
>    for $j = 1, \ldots, c$
>        if (distance($P_s$,$c_j$)<distance($P_s$, $c_i$)
>        then
>            return(false);
>    return(true); }

$\diamond$

When we have calculated a list with all the relevant hyperplanes, we still need to construct the multilayer perceptron out of these values. See section 9.2.1 and 9.2.2 for details.

**Algorithm 20 (Constructing the MLP)**

> Construct_MLP(List_of_Relevant_Hyperplanes)
> {
>    $l :=$ length of List_of_Relevant_Hyperplanes;
>    for $t = 1, \ldots, l$
>        for $s = 1, \ldots, n$
>            $W_1(s, t) :=$ normalvector($H_t$)$[s]$;
>        $\Theta_1[t] :=$ abscissa($H_t$);
>    for $t = 1, \ldots, c$;
>        for $s = 1, \ldots, l$;
>            $W_2(s, t) := 1$ with $H_s = H_{tj}$;
>            $W_2(s, t) := 1$ with $H_s = H_{it}$;
>        $\Theta_2[t] := 0.5-$ number of $s$ with $W_2(t, s) = 1$;
>    for $t = 1, \ldots,$ number of classes;
>        for $s = 1, \ldots, c$;
>            if ($c_s \in \mathcal{C}_t$)
>            then
>                $W_3(s, t) := 1$
>            else
>                $W_3(s, t) := 0$
>        $\Theta_3[t] := -0.5$
> }

$\diamond$

# B.4 Deriving a Multilayer Perceptron with Continuous Output

Our method of deriving an MLP from the information that fuzzy clustering provides, was designed for classification problems. But this method can be expanded to problems where a continuous output is desired.

In chapter 10 we have described such a problem. The input data where clustered, and afterwards an output was assigned to each of the clusters. An MLP would model this in the only case that only the output unit with the highest activation would provide the output.

In practice this looks different. Each of the output units has an activation value, while the unit which represents the cluster containing the input value has the biggest activation.

To achieve exactly the desired output for our prototypes, we use a system of linear equations to calculate the activation of a single output unit that is combining the previous output units. This provides the MLP with a good starting point for further learning.

For the below algorithm we assume that the logistic function $f(x) = \frac{1}{1+e^{-x}}$ is used. Output consist of the output values for each prototype that are assigned to the clusters.

**Algorithm 21 (Continuous output)**

*void Construct_Output_Neuron(Prototypes $c_1, \ldots, c_c$,*
*        MultilayerPerceptron M, Vector Output)*
*{*
*    VectorArray Results;*
*    Vector m,s; for $i = 1, \ldots, c$*
*        $m[i] := \ln(\frac{1}{Output[i]} - 1)$; Results[i] := M(c_i);*
*    Vector $s := Results^{-1} \cdot m$;*
*    for $i = 1, \ldots, c$*
*        $W(u_i, o) := s[i]$;*
*}*

$\diamond$

**Remark 31**
*Note that this algorithm ist constructed for normalized output data. If we do not have this, we have to normalize the dataset, use the same values to scale the output and then use the given construction mehtod. Afterwards the output of the MLP can be transformed back to original scale.*

# Appendix C

# Analysis of the weather data

## C.1 The prototypes of the clustering

Here you find the data evaluated by [rehm03] that we used in chapter 10 for the calculations.

| Cluster | Temp-erature | Air pressure | Visi-bility | Head-wind | Side-wind | South-wind |
|---------|--------------|--------------|-------------|-----------|-----------|------------|
| 1 | 14,04 | 1013,48 | 29109,91 | 10,24 | 7,88 | 10,75 |
| 2 | 10,97 | 1010,77 | 26372,87 | 3,74 | 3,59 | 3,02 |
| 3 | 6,17 | 1026,51 | 47304,94 | 4,35 | 3,2 | -2,03 |
| 4 | 21,57 | 1015,2 | 41237,23 | 9,69 | 2,49 | 2,91 |
| 5 | 22,93 | 1017,41 | 30166,4 | 1,62 | 1,96 | -0,12 |
| 6 | 5,66 | 1030,2 | 16079,51 | 4,54 | 3,03 | 0,56 |
| 7 | 7,43 | 1018,92 | 16951,56 | -0,79 | 1,33 | -0,99 |
| 8 | 20,08 | 1018,87 | 31777,27 | 5,49 | 6,19 | -7,1 |

Table C.1: The prototypes for cloudless weather in 1998

| Cluster | Temp-erature | Air pressure | Visi-bility | Head-wind | Side-wind | South-wind |
|---|---|---|---|---|---|---|
| 1 | 14,04 | 1013,48 | 29109,91 | 10,24 | 7,88 | 10,75 |
| 2 | 10,97 | 1010,77 | 26372,87 | 3,74 | 3,59 | 3,02 |
| 3 | 6,17 | 1026,51 | 47304,94 | 4,35 | 3,2 | -2,03 |
| 4 | 21,57 | 1015,2 | 41237,23 | 9,69 | 2,49 | 2,91 |
| 5 | 22,93 | 1017,41 | 30166,4 | 1,62 | 1,96 | -0,12 |
| 6 | 5,66 | 1030,2 | 16079,51 | 4,54 | 3,03 | 0,56 |
| 7 | 7,43 | 1018,92 | 16951,56 | -0,79 | 1,33 | -0,99 |
| 8 | 20,08 | 1018,87 | 31777,27 | 5,49 | 6,19 | -7,1 |

Table C.2: The prototypes for cloudless weather in 1999

| Cluster | Temp-era-ture | Clouds 1 | 2 | Air pres-sure | Visi-bility | Cloud Layer 1 | 2 | Head-wind | Side-wind | South-wind |
|---|---|---|---|---|---|---|---|---|---|---|
| 1 | 7,97 | 3,37 | 4,06 | 1016,66 | 12548,65 | 18,43 | 42,66 | 5,1 | 4,42 | 4,04 |
| 2 | 12,49 | 2,05 | 3,12 | 1009,77 | 12965,58 | 12,67 | 31,77 | 3,89 | 3,59 | 2,67 |
| 3 | 4,69 | 2,08 | 3,79 | 1021,08 | 9705,55 | 12,45 | 31,52 | 3,53 | 2,95 | 2,35 |
| 4 | 16,52 | 2,13 | 3,52 | 1014,77 | 29127,86 | 33,23 | 69,07 | 4,91 | 3,36 | 1,71 |
| 5 | 13,42 | 2,23 | 3,7 | 1014,09 | 25703,55 | 40,62 | 245,68 | 4,86 | 3,25 | 2,46 |
| 6 | 6,74 | 2,09 | 3,64 | 1022,03 | 21926,3 | 20,22 | 36,26 | 5,75 | 6,72 | -6,21 |
| 7 | 12,16 | 2,13 | 3,34 | 1009,1 | 35328,93 | 26,34 | 51,13 | 12,89 | 4,43 | 6,74 |
| 8 | 10,59 | 2,12 | 3,71 | 1008,23 | 20356,5 | 17,57 | 35,97 | 9,24 | 8,8 | 11,24 |

Table C.3: The prototypes for clouded weather in 1998

| Cluster | Temp-era-ture | Clouds 1 | 2 | Air pres-sure | Visi-bility | Cloud Layer 1 | 2 | Head-wind | Side-wind | South-wind |
|---|---|---|---|---|---|---|---|---|---|---|
| 1 | 16,49 | 2,2 | 3,64 | 1015,18 | 24401,28 | 37,36 | 244,22 | 4,88 | 3,85 | 2,2 |
| 2 | 6,48 | 2,14 | 3,78 | 1019,66 | 10261,69 | 12,73 | 39,12 | 3 | 3,29 | 1,13 |
| 3 | 10,9 | 2,11 | 3,61 | 1008,65 | 23042,18 | 17,78 | 37,01 | 8,57 | 9,12 | 11,12 |
| 4 | 15,49 | 2,13 | 3,45 | 1018,31 | 28358,36 | 34,01 | 69,65 | 3,95 | 5,55 | -4,68 |
| 5 | 16,1 | 2,09 | 2,72 | 1012,54 | 15623,37 | 17,13 | 44,74 | 3,95 | 3,59 | 2,23 |
| 6 | 4,82 | 2,05 | 3,36 | 1004,08 | 16952,03 | 12,23 | 27,37 | 8,55 | 4,55 | 3,29 |
| 7 | 16,96 | 2,19 | 3,32 | 1010,57 | 35863,48 | 32,4 | 79,41 | 9,2 | 4,3 | 6,15 |
| 8 | 7,68 | 3,45 | 4,1 | 1011,3 | 13632,72 | 17,36 | 50,19 | 6,21 | 5,7 | 4,83 |

Table C.4: The prototypes for clouded weather in 1999

# List of Figures

# List of Tables

# Symbols

# Abbreviations

# Algorithms

# Bibliography

[1] D.E. Anderson and L.O. Hall: Mr. FIS: Mamdani Rules Fuzzy Inference System. IEEE SMC'99 conference, Vol.V, (1999),38-243

[2] N. Ashford, H.P. Stanton, C.A. Moore: Airport Operation, McGraw-Hill Professionals, New York (1996)

[3] P. Bachmann: Internationale Flughäfen im Vergleich (in German), Motorbuch Verlag, Stuttgart (1997)

[4] S.A. Bailey, Y.-H. Chen: A Two Layer Network using the OR/AND Neuron. IEEE Fuzzy Systems Conference, Vol. II, (19998), 1211-1216

[5] P. Bauer, E.P. Klement, A. Leikermoser, B. Moser: Interpolation and approximation of real input-output functions using fuzzy rule bases. In: [55], 245-254

[6] O.E. Barndorff-Nielsen, J.L. Jensen, W.S. Kendall (Ed.): Networks and Chaos - Statistical and Propbabilistic Aspects, Chapman & Hall, (1993)

[7] J.C. Bezdek: Pattern Recognition with Fuzzy Objective Function Algorithms, Plenum Press, New York (1981)

[8] M. Berthold, D.J. Hand (Ed.): Intelligent Data Analysis, Springer, Heidelberg (1999)

[9] M.R. Berthold, K.-P. Huber: Constructing Fuzzy Graphs from Examples. Intelligent Data Analysis 3, Elsevier, (1999), 37-53

[10] C.B. Barber, D.P. Dobkin. H. Huhdanpaa: The Quickhull Algorithm for Convex Hulls, ACM Transactions on Mathematical Software, vol. 22, no. 4, p.469–483, (1996)

[11] L. Breiman, J.H. Freidman, R.A. Olshen, C.J. Stone: Classification and Regression Trees, Chapman & Hall, New York (1984)

[12] C.J.C. Burges: A Tutorial on Support Vector Machines for Pattern Recognition, Data Mining and Knowledge Discovery, Vol. 2, Boston(1998), 121-167

[13] D.Butnariu, E.P. Klement: Triangular Norm-Based Measures and Games with Fuzzy Coalitions, Kluwer Academic Publishers, Dordrecht (1993)

[14] J.L. Castro, C.J. Mantras, J.M. Benitez: Multilayer Neural Networks as Fuzzy Ruls Systems, Proceedings of the 8th International Conference IPMU, Vol.III, Madrid (200), 1822-1829

[15] J.L. Castro, E. Trillas, S. Cubillo: On Consequence in Approximate Reasoning. Journal of Applied Non-Classical Logics **4** (1994), 91-103

[16] M.-Y. Chen, D.A. Linkens: Fuzzy Modelling Approach Using Clustering Neural Networks. In: IEEE Fuzzy Systems Conference 1998, Vol II, 1088-1093

[17] O. Cordón, M. José del Jesus, F. Herrera: Analysing the Reasoning Mechanism in Fuzzy Rule Based Classification Systems. Mathware & Soft Computing 5 (1998), 321-332

[18] R.N. Davé: Characterisation and Detection of Noise in Clustering. Pattern Recognition Letters 12 (1991), 657-664

[19] R.N. Davé, R. Krishnapuram: Robust Clustering Methods: A Unified View, IEEE Transactions on Fuzzy Systems 5 (1997), 270-293

[20] M. Demirci: Fuzzy Functions and their Fundamental properties. Fuzzy Sets and Systems, v.106 n.2, (1999), 239-246

[21] G.L. Donohue, A.G. Zellweger, H. Rediess, Ch. Push (Ed.): Air Transportation Systems Engineering, American Institue of Aeronautics and Astronautics, Inc., Virginia (2001)

[22] D. Dubois, H. Prade: An Introduction to Possibilistic and Fuzzy Logics. In: [99]

[23] W. Duch, R. Adamczak, K. Grabczewski: A New Methodology of Extraction, Optimization and Application of Crisp and Fuzzy Logical Rules. IEEE Transactions on Neural Networks, Vol.12, No.2, (2001), 277-306

[24] R.A. Fisher: The Use of Multiple Measurements in Taxonomic Problems. Annals of Eugenics 7 (1936), 179-188

[25] S. Fortune: Voronoi Diagrams and Delaunay Triangulations, in: D.-Z. Du, F. Hwang (Ed.): Computing in Euclidean Geometry, World Scientific, Lecture Notes Series on Computing – Vol. 1, Singapore (1992), p. 193-233n

[26] E. Frías-Martínez, J. Gutiérrez-Ríos, F. Fernández-Hernández: Rule-Driven Architechture for a Łukasiewicz T-norm, Proceedings of the 8th International Conference IPMU Information Processing and Management of Uncertainty in Knowledge-based Systems, Vol. I, Madrid (2000), 25-29

[27] X. Fron, EUROCONTROL: Performance Review in Europe, in [21], p.49-60

[28] X. Fu, L. Wang: Data Dimensionality Reduction With Application to Simplifying RBF Network Structure and Improving Classification Performance, IEEE Transactions on Systems, Man and Cybernetics, Part B: Cybernetics, Vol. 33, No 3, June 2003

[29] B. Gabrys, A. Bargiela: General Fuzzy Min-Max Neural Network for clsutering and Classfication. IEEE Transactions on Neural Networks, Vol.11, No.3, (2000), 769-783

[30] M. Gallagher, T. Downs: Visualization of Learning in Multilayer Perceptron Networks Using Principal Component Analysis. IEEE Transactions on Systems, Man and Cybernetics. Vol.33, No.1, (2003). 28-34

[31] H. Genther, M. Glesner: Automatic Generation of a Fuzzy Classification System using Fuzzy Clustering Methods. Proc. ACM Symposium on Applied Computing (SAC'94), Phoenix (1994), 180-183

[32] D. Gustafson, W. Kessel: Fuzzy Clustering with a Fuzzy Covariance Matrix. Proc. IEEE CDC, San Diego (1979), 761-766

[33] I. Hayashi,T. Maeda, A. Bastian, L.C. Jain: Generation of Fuzzy Decision Trees by Fuzzy ID3 with Adjusting Mechanism of AND/OR Operators. IEEE Fuzzy Systems Conference 1998, I, 681-685

[34] U. Höhle: On the Fundamentals of Fuzzy Set Theory. In: Journal of Mathematical Analysis and Applications 201 (1996), 786-826

[35] U. Höhle: Many-valued Equalities, Singletons and Fuzzy Partitions. In: Soft Computing 2, Springer (1998), 134-140

[36] F. Höppner, F. Klawonn: A New Approach to Fuzzy Partitioning. In NAFIPS01, Vancouver(2001), 1419–1424

[37] F. Höppner, F. Klawonn, R. Kruse, T. Runkler: Fuzzy Cluster Analysis. Wiley, Chichester (1999)

[38] H. Ishibuchi: A Fuzzy Classifier System that Generates Linguistic Rules for Pattern Classification Problems. In: Fuzzy Logic, Neural Networks, and Evolutionary Computation, Springer, Berlin (1996), 35-54

[39] C.Z. Janikow: Fuzzy Decision Trees: Issues and Methods, IEE Transactions on Systems, Man and Cybernetics - Part B: Cybernetics, Vol. 28, No. 1, (1998), 1-14

[40] T. Joachims: Support Vector Machines (in German), Künstliche Intelligenz, 4/99, Bremen (1999), 54-55

[41] A. Keller: Objective Function Based Fuzzy Clustering in Air Traffic Management, Magdeburg (2002)

[42] A. Keller, F. Klawonn: Fuzzy Clustering with Weighting of Data Variables. International Journal of Uncertainty, Fuzziness and Knowledge-Based Systems 8 (2000), 735-746

[43] A. Keller, F. Klawonn: Adaption of Cluster Sizes in Objective-Function Based Fuzzy Clustering. In: T.Leondes (ed.): Intelligent Systems: Techiques and Applications - Database and Learning Systems, volume IV, CRC Press (2002), 181-199

[44] F. Klawonn: Fuzzy Sets and Similarity-Based Reasoning, Braunschweig (1996)

[45] F. Klawonn: Visualisierung von Fuzzy-Clusteranalyse-Ergebnissen (in German), Proceedings 12. Workshop Fuzzy Systeme, Dortmund (2002), 1-12

[46] F. Klawonn, E.P. Klement: Mathematical Analysis of Fuzzy Classifiers. In: Mathematical analysis of fuzzy classifiers. In: X. Liu, P. Cohen, M. Berthold (eds.): Advances in intelligent data analysis. Springer, Berlin (1997), 359-370

[47] F. Klawonn, R. Kruse: Derivation of Fuzzy Classification Rules from Multidimensional Data. In: G.E. Lasker, X. Liu (eds.): Advances in intelligent data analysis. The International Institute for Advanced Studies in Systems Research and Cybernetics, Windsor, Ontario (1995), 90-94

[48] A. Klose, A. Nürnberger: Applying Boolean Transformations to Fuzzy Rule Bases. In: Proc. 7th European Congress on Intelligent Techniques and Soft Computing (EUFIT'99), Verlag Mainz, Aachen (1999)

[49] A. Klose, A. Nürnberger, D. Nauck, R. Kruse: Data Mining with Neuro-Fuzzy Models. In: A. Kandel, M. Last, H. Burka (ed.): Data Mining and Computational Intelligence, Physica, Heidelberg (2001)

[50] B. Kosko: Fuzzy Systems as Universal Approximators. Proc. IEEE International Conference on Fuzzy Systems 1992, San Diego (1992), 1153-1162

[51] A. Klose, A. Nürnberger, D. Nauck: Some Approaches to Improve the Interpretability of Neuro-Fuzzy Classifiers. In: Proc. 6th European Congress on Intelligent Techniques and Soft Computing (EUFIT'98), Aachen (1998), 629-633

[52] A. Klose: Partially Supervised Learning of Fuzzy Classification Rules, Magdeburg (2004)

[53] R. Krishnapuram, J. Keller: A Possibilistic Approach to Clustering. IEEE Transactions on Fuzzy Systems 1 (1993), 98-110

[54] R. Kruse, J. Gebhardt, F. Klawonn: Foundations of Fuzzy Systems. Wiley, Chichester (1994)

[55] R. Kruse, J. Gebhardt, R. Palm (eds.): Fuzzy Systems in Computer Science. Vieweg, Braunschweig (1994)

[56] M. Kubat: Decision Trees Can Initialize Tadial-Basis Function Networks, IEEE Transactions on Neural Networks, Vol. 9, No. 5, (1998), 813-821

[57] L.I. Kuncheva: How good are Fuzzy If-Then Classifiers? IEEE Transactions on Systems, Man and Cybernetics 30, Part B (2000), 501-509

[58] L.I. Kuncheva: Fuzzy Classifier Design, Springer, Heidelberg (2000)

[59] L.I. Kuncheva:'Fuzzy' vs 'Non-fuzzy' in Combining Classifiers: an experimental study. Proc LFA'01, Mons, Belgium(2001), 11-22

[60] L.I. Kuncheva: Switching Between Selection and Fusion in Combining Classifiers: An Experiment. IEEE TRansactions on Systems, Man, and Cyberentics, Vol.32, No.2, (2002), 146-156

[61] L.I. Kuncheva, C.J. Whitaker: Feature Subsets for Classifier Combination: An Enumerative Experiment. MCS, Lecture Notes in Computer Science, LNCS 2096, Camebridge (2001), 228-237

[62] J. Lee, S. Chae: Analysis on Function Duplicating Capabilities of Fuzzy Controllers. Fuzzy Sets and Systems **56** (1993), 127-143

[63] H.-X. Li, C.L.P. Chen: The Equivalence Between Fuzzy Logic Systems and Feedforward Neural Networks. IEEE Transactions on Neural Networks, vol.11, No.2, (2000)

[64] J.W.Lloyd: Foundations of Logic Programming. Springer, Berlin (1987)

[65] E.H. Mamdani, S. Assilian: An Experiment in Linguistic Synthesis with a Fuzzy Logic Controller. Intern. Journ. of Man Machine Studies 8 (1975), 1-13

[66] E. McCluskey: Algebraic Minimization and the Design of Two-terminal Contact Networks, Bell System Technical Journal, vol. 35, (1956), 1417-1444

[67] B.J.A. Mertens, D.J. Hand: Adjusted Estimation for the Combiantion of Classifiers. In: D.J. Hand, J.N. Kok, M.R. Berthold (eds.): IDA'99, LNCS, Springer, Heidelberg (1999), 317-330

[68] K.D. Meyer Gramann: Fuzzy Classification: An Overview. In: [55], 277-294

[69] S. Mitra: Self-Organizing Neural Networks as a Fuzzy Classifier, IEEE Transactiosn on Systems, Man and Cybernetics, Vol.24, No.3, (1994), 385-399

[70] S. Mitra, S.K. Pal, M.K. Kundu: Fingerprint Classification using a Fuzzy Multilyer Perceptron, Neural comput & Applic, Springer, London(1994), 2:227-233

[71] S. Mitra, S.K. Pal: Logical Operation Based Fuzzy MLP for Classification and Rule Generation, Neural Networks, Vol7, No2, Elsevier, (1994), 353-373

[72] S. Mitra, R.K. De, S.J. Pal: Knowledge-Based Fuzzy MLP for Classification and Rule Generation, IEEE TRansactions on Neural Networks, Vol.8, No.6, (1997), 1338-1350

[73] S. Mitra: Neuro-Fuzzy Rule Generation: Survey in Soft Computing Framework, IEEE TRansactions on Neural Networks, Vol.11, No.3, (2002), 748-768

[74] B. Moser: Sugeno Controllers with a Bounded Number of Rules are Nowhere Dense. Fuzzy Sets and Systems 104 (1999), 269-277

[75] D. Nauck, R. Kruse: NEFCLASS – A Neuro-fuzzy Approach for the Classification of Data. In: K.M. George, J.H. Carrol, E. Deaton, D. Oppenheim, J. Hightower (eds.): Applied Computing 1995: Proc. of the 1995 ACM Symposium on Applied Computing. ACM Press, New York (1995), 461-465

[76] D.Nauck, F.Klawonn, R.Kruse: "Foundations of Neuro-Fuzzy Systems", Wiley, Chichester (1997)

[77] D.Nauck, F.Klawonn, R.Kruse: Neuronale Netze und Fuzzy-Systeme (in german), Vieweg, Braunschweig (1996)

[78] R.L. Neufville, A.R. Odoni, R. De Neufville: Airport Systems: Planning, Design and Management, McGraw-Hill, New York (2002)

[79] A. Nürnberger, A. Klose, R. Kruse: Discussing Cluster Shapes of Fuzzy Classifiers. Proc. 18th Conf. of the North American Fuzzy Information Processing Society (NAFIPS'99), New York (1999), 546-550

[80] A. Nürnberger, A. Klose, R. Kruse: Analyzing Borders between partially Contradicting Fuzzy Classification Rules. Proc. 19th Conf. of the North American Fuzzy Information Processing Society (NAFIPS'00), Atlanta (2000), 59-63

[81] N.P. Pal, S. Chakraborty: Fuzzy Rules From ID3-Type Decision Trees for Real Data. IEEE TRansactions on Systems, Man and Cybernetics, Part B: Cybernetics, Vol.31, No.5, (2001), 745-754

[82] S.K. Pal, S. Mitra: Fuzzy Versions of Kohonen's Net and MLP Classification: Performance Evaluation for Certain Nonconvex Decision Regions, Informations Sciences 76, (1994), 297-337

[83] W. Pedrycz: Algorithms of Fuzzy Clustering with partial Supervision. Pattern Recognition Letters 23 (1985), 13-20

[84] Performany Review Commision: Performance Review Report 7, An Assessment of Air Traffic Management in Europe during the Calender Year 2003, EUROCONTROL, Brussels (2004)

[85] W. Quine: A Way to Simplify Truth Functions. American Mathematical Society, 62, (1955), 627-631.

[86] J.R. Quinlan: Induction of Decision Trees, Machine Learning, 1, Boston (1986), 81-106

[87] F. Rehm: Data Mining Methoden zur Bestimmung des Einflusees von Wetterfaktoren auf Anflugverspätungen an Flughäfen (in German), Braunschweig (2003)

[88] B.D. Ripley: Neural Networks and Related Methods for Classification. J.R. Statist. Coc. B, 56, No.3, (1994), 409-456

[89] B.D. Ripley: Statistical Aspects of Neural Networks. In [6] Chapman & Hall, (1993)

[90] D.E. Rumelhart, G.E. Hinton and R.J.Williams: Learning Internal Representation by Error Propagation. In [91], MIT Press, Cambridge (1986), 318-362

[91] D.E. Rumelhart and R.J.McClelland (eds.): Parallel istributed Processing: Explorations in the Microstructures of Cognition. Foudations, Vol. 1, MIT Press, Cambridge (1986)

[92] T.P. Ryan: Modern Regression Methods, Wiley, New York (1997)

[93] B. von Schmidt: Geometrische Veranschaulichung von Fuzzy-Klassifikationssystemen und Multilayer Perceptrons und Parallelen zwischen beiden Modellen (in German), 12. Workshop Fuzzy Systeme, Karlsruhe (2002), 58-69

[94] B. von Schmidt, F. Klawonn: Fuzzy Max-Min Classifiers decide locally on the Basis of Two Attributes. Mathware and Soft Computing 6 (1999), 91-108

[95] B. von Schmidt, F. Klawonn: Construction of Fuzzy Classification Systems with the Łukasiewicz-t-norm. Proc. 19th Conf. of the North American Fuzzy Information Processing Society (NAFIPS'00), Atlanta (2000), 109-113

[96] B. von Schmidt, F. Klawonn: Construction of a Multilayer Perceptron for a Piecewise Linearly Separable Classification Problem. Proc.
Joint 9th International Fuzzy Systems Association World Congress and
20th North American Fuzzy Information Processing Society International Conference, Vancouver (2001)

[97] B. von Schmidt, F. Klawonn: Extracting Fuzzy Classification Rules from
Fuzzy Clusters on the Basis of Separating Hyperplanes. In: J. Casillas,
O. Cordón, F. Herrera, L- Magdalena (eds.): Interpretability issues in
fuzzy modelling. Springer, Berlin (2003), 621-643

[98] R. Schmitz: IST-Analyse des Flughafen Frankfurt (FRA-IST), Teil2:
Luftseitiger Verkehr (In German), German Aerospace Center, Institut
of Flight Guidance, Braunschweig(2001)

[99] P. Smets, E.H. Mamdani, D. Dubois, H. Prade (eds.): Non-Standard
Logics for Automated Reasoning. Academia Press, London (1988)

[100] M. Sugeno, T. Yasukawa: A Fuzzy Logic-based Approach to Qualita-
tive Modelling. IEEE Transactions on Fuzzy Systems 1 (1993), 7-31

[101] H. Takagi, M. Sugeno: Fuzzy Identification of Systems and its Application to Modelling and Control. IEEE Trans. on Systems, Man, and
Cybernetics 15 (1985), 116-132

[102] S.L. Tanimoto: The Elements of Artificial Intelligencec - An Introduction using Lisp, Computer Science Press, (1987)

[103] A.B. Tickle, R. Andrews, M. Golea, J. Diedrich: The Truth will Come
to Light: directions and Challenges in Extracting the Knowledge Em-
bedded Within Trained Artificial Neural Networks. IEEE Transactions
on Neural Networks, Vol.9, No.6, (1998), 1057-1068

[104] H. Timm: Fuzzy Clusteranalyse: Methoden zur Exporaltion von Daten
mit fehlenden Werten sowie klassifizierten Daten (in German), Magdeburg (2002)

[105] L.X. Wang: Fuzzy Systems are Universal Approximators. Proc. IEEE
International Conference on Fuzzy Systems 1992, San Diego (1992),
1163-1169

[106] R. Weber: Fuzzy-ID3: A Class of Methods for Automatic Knowledge
Acquisition. In: Proc. 2nd International Conference on Fuzzy Logic and
Neural Networks, Iizuka (1992), 265-268

[107] L. Zadeh: Fuzzy Sets, In: Information Control, 8:338-353, 1965

[108] A. Zell: Simulations Neuronaler Netze. Addison-Wesley, Bonn (1994)

# Index